Other Titles in This Series

Volume

7   **Andy R. Magid**
Lectures on differential Galois theory
1994

6   **Dusa McDuff and Dietmar Salamon**
J-holomorphic curves and quantum cohomology
1994

5   **V. I. Arnold**
Topological invariants of plane curves and caustics
1994

4   **David M. Goldschmidt**
Group characters, symmetric functions, and the Hecke algebra
1993

3   **A. N. Varchenko and P. I. Etingof**
Why the boundary of a round drop becomes a curve of order four
1992

2   **Fritz John**
Nonlinear wave equations, formation of singularities
1990

1   **Michael H. Freedman and Feng Luo**
Selected applications of geometry to low-dimensional topology
1989

UNIVERSITY LECTURE SERIES VOLUME 6

J-Holomorphic Curves and Quantum Cohomology

Dusa McDuff
Dietmar Salamon

AMERICAN MATHEMATICAL SOCIETY
Providence, Rhode Island

1991 *Mathematics Subject Classification.* Primary 53C15; Secondary 58F05, 57R57.

Library of Congress Cataloging-in-Publication Data

McDuff, Dusa, 1945–
    J-holomorphic curves and quantum cohomology/Dusa McDuff, Dietmar Salamon.
        p.   cm. — (University lecture series, ISSN 1047-3998; v. 6)
    Includes bibliographical references and indexes.
    ISBN 0-8218-0332-8
    1. Symplectic manifolds.   2. Holomorphic functions.   3. Homology theory.   I. Salamon, D. (Dietmar) II. Title.   III. Series: University lecture series (Providence, R.I.); v. 6.
QA649.M42   1994
516.3′62—dc20   94-25414
                                                                                                        CIP

Contents

1 Introduction   **1**
   1.1   Symplectic manifolds 1
   1.2 J-holomorphic curves 3
   1.3   Moduli spaces 4
   1.4   Compactness 5
   1.5   Evaluation maps 6
   1.6   The Gromov-Witten invariants 8
   1.7   Quantum cohomology 9
   1.8   Novikov rings and Floer homology 11

2 Local Behaviour   **13**
   2.1   The generalised Cauchy-Riemann equation 13
   2.2   Critical points 15
   2.3   Somewhere injective curves 18

3 Moduli Spaces and Transversality   **23**
   3.1   The main theorems 23
   3.2   Elliptic regularity 25
   3.3   Implicit function theorem 27
   3.4   Transversality 33
   3.5   A regularity criterion 38

4 Compactness   **41**
   4.1   Energy 42
   4.2   Removal of Singularities 43
   4.3   Bubbling 46
   4.4   Gromov compactness 50
   4.5   Proof of Gromov compactness 52

5 Compactification of Moduli Spaces   **59**
   5.1   Semi-positivity 59
   5.2   The image of the evaluation map 62
   5.3   The image of the p-fold evaluation map 65
   5.4   The evaluation map for marked curves 66

6 Evaluation Maps and Transversality   **71**
   6.1 Evaluation maps are submersions 71
   6.2 Moduli spaces of N-tuples of curves 74
   6.3 Moduli spaces of cusp-curves 75
   6.4 Evaluation maps for cusp-curves 79
   6.5 Proofs of the theorems in Sections 5.2 and 5.3 81
   6.6 Proof of the theorem in Section 5.4 83

7 Gromov-Witten Invariants   **89**
   7.1 Pseudo-cycles 90
   7.2 The invariant Φ 93
   7.3 Examples 98
   7.4 The invariant Ψ 101

8 Quantum Cohomology   **107**
   8.1 Witten's deformed cohomology ring 107
   8.2 Associativity and composition rules 114
   8.3 Flag manifolds 119
   8.4 Grassmannians 123
   8.5 The Gromov-Witten potential 131

9 Novikov Rings and Calabi-Yau Manifolds   **141**
   9.1 Multiply-covered curves 142
   9.2 Novikov rings 144
   9.3 Calabi-Yau manifolds 148

10 Floer Homology   **153**
   10.1 Floer's cochain complex 153
   10.2 Ring structure 159
   10.3 A comparison theorem 160
   10.4 Donaldson's quantum category 162
   10.5 Closing remark 165

A Gluing   **167**
   A.1 Cutoff functions 168
   A.2 Connected sums of J-holomorphic curves 170
   A.3 Weighted norms 171
   A.4 An estimate for the inverse 173
   A.5 Gluing 176

B Elliptic Regularity   **181**
   B.1 Sobolev spaces 181
   B.2 The Calderon-Zygmund inequality 185
   B.3 Cauchy-Riemann operators 190
   B.4 Elliptic bootstrapping 192

   Bibliography   **197**

   Indexes   **203**

   Index of Notations   **209**

Preface

The theory of J-holomorphic curves has been of great importance to symplectic topologists ever since its inception in Gromov's paper [26] of 1985. Its applications include many key results in symplectic topology: see, for example, Gromov [26], McDuff [42, 45], Lalonde–McDuff [36], and the collection of articles in Audin–Lafontaine [5]. It was also one of the main inspirations for the creation of Floer homology [18, 19, 73], and recently has caught the attention of mathematical physicists through the theory of quantum cohomology: see Vafa [82] and Aspinwall–Morrison [2].

Because of this increased interest on the part of the wider mathematical community, it is a good time to write an expository account of the field, which explains the main technical steps in the theory. Although all the details are available in the literature in some form or other, they are rather scattered. Also, some improvements in exposition are now possible. Our account is not, of course, complete, but it is written with a fair amount of analytic detail, and should serve as a useful introduction to the subject. We develop the theory of the Gromov-Witten invariants as formulated by Ruan in [64] and give a detailed account of their applications to quantum cohomology. In particular, we give a new proof of Ruan–Tian's theorem [67, 68] that the quantum cup-product is associative.

Many people have made useful comments which have added significantly to our understanding. In particular, we wish to thank Givental for explaining quantum cohomology, Ruan for several useful discussions and for pointing out to us the connection between associativity of quantum multiplication and the WDVV-equation, Taubes for his elegant contribution to Section 3.4, and especially Gang Liu for pointing out a significant gap in an earlier version of the gluing argument. We are also grateful to Lalonde for making helpful comments on a first draft of this manuscript. The first author wishes to acknowledge the hospitality of the University of California at Berkeley, and the grant GER-9350075 under the NSF Visiting Professorship for Women program which provided partial support during some of the work on this book.

Dusa McDuff,
Mathematics Department
SUNY at Stony Brook,
Stony Brook, NY 11794, USA
dusa@math.sunysb.edu

Dietmar Salamon,
Mathematics Institute,
University of Warwick,
Coventry CV4 7AL, Great Britain
das@maths.warwick.ac.uk

Chapter 1

Introduction

The theory of J-holomorphic curves is one of the new techniques which have recently revolutionized the study of symplectic geometry, making it possible to study the global structure of symplectic manifolds. The methods are also of interest in the study of Kähler manifolds, since often when one studies properties of holomorphic curves in such manifolds it is necessary to perturb the complex structure to be generic. The effect of this is to ensure that one is looking at persistent rather than accidental features of these curves. However, the perturbed structure may no longer be integrable, and so again one is led to the study of curves which are holomorphic with respect to some non-integrable almost complex structure J.

The present book has two aims. The first is to establish the fundamental theorems in the subject, and in particular the existence of the Gromov-Witten invariants. The second is to serve as an introduction to the subject. These two aims are, of course, somewhat in conflict, and in different parts of the book different aspects are predominant. The book is written in logical order. Chapters 2 through 6 establish the foundational Fredholm theory and compactness results needed to set up the theory. Chapters 7 and 8 then discuss the Gromov-Witten invariants, and their application to quantum cohomology, for symplectic manifolds which satisfy a certain positivity (or monotonicity) condition. Chapter 9 treats extensions to Calabi-Yau manifolds, and Chapter 10 relations with Floer homology. Appendix A develops a gluing technique for for J-holomorphic spheres, which is an essential ingredient of the proof of the associativity of quantum multiplication. Finally, Appendix B reviews and proves some basic facts on elliptic regularity.

It may be best not to read the book in chronological order, but rather to start with Chapter 7. The present introductory chapter aims to outline enough of the theory of J-holomorphic curves to make this approach feasible. We assume that the reader is familiar with the elements of symplectic geometry. Good references are McDuff–Salamon [52] and the introductory articles in Audin–Lafontaine [5].

1.1 Symplectic manifolds

A symplectic structure on a smooth $2n$-dimensional manifold M is a closed 2-form ω which is non-degenerate in the sense that the top-dimensional form ω^n does not vanish. By Darboux's theorem, all symplectic forms are locally diffeomorphic to the standard linear form

$$\omega_0 = dx_1 \wedge dx_2 + \ldots + dx_{2n-1} \wedge dx_{2n}$$

on Euclidean space $\mathbb{R}^{2n}$. This makes it hard to get a handle on the global structure of symplectic manifolds. Variational techniques have been developed which allow one to investigate some global questions in Euclidean space and in manifolds such as cotangent bundles which have some linear structure: see [52] and the references contained therein. But the method which applies to the widest variety of symplectic manifolds is that of J-holomorphic curves.

Here J is an almost complex structure on M which is tamed by ω. An almost complex structure is an automorphism J of the tangent bundle TM of M which satisfies the identity $J^2 = -\mathbb{1}$. Thus J can be thought of as multiplication by i, and it makes TM into a complex vector bundle of dimension n. The form ω is said to **tame** J if

$$\omega(v, Jv) > 0$$

for all nonzero $v \in TM$. Geometrically, this means that ω restricts to a positive form on each complex line $L = \mathrm{span}\{v, Jv\}$ in the tangent space $T_x M$. Given ω the set $\mathcal{J}_\tau(M, \omega)$ of almost complex structures tamed by ω is always non-empty and contractible. Note that it is very easy to construct and perturb tame almost complex structures, because they are defined by pointwise conditions. Note also that, because $\mathcal{J}_\tau(M, \omega)$ is path-connected, different choices of $J \in \mathcal{J}_\tau(M, \omega)$ give rise to isomorphic complex vector bundles (TM, J). Thus the Chern classes of these bundles are independent of the choice of J and will be denoted by $c_i(M)$.

In what follows we shall only need to use the first Chern class, and what will be relevant is the value which it takes on embedded 2-spheres in M. If S is such a sphere in the homology class $A \in H_2(M)$, we will need to calculate $c_1(A)$. This is just the first Chern class of the restriction of the bundle TM to S. But every complex bundle E over a 2-sphere S decomposes as a sum of complex line bundles $L_1 \oplus \ldots \oplus L_n$. Correspondingly

$$c_1(E) = \sum_i c_1(L_i).$$

Since the first Chern class of a complex line bundle is just the same as its Euler class, it is often easy to calculate the $c_1(L_i)$ directly. For example, if A is the class of the sphere $S = \mathrm{pt} \times S^2$ in $M = V \times S^2$ then it is easy to see that

$$TM\big|_S = TS \oplus L_2 \oplus \ldots \oplus L_n,$$

where the line bundles L_k are trivial. It follows that

$$c_1(A) = c_1(TM\big|_S) = c_1(TS) = \chi(S) = 2$$

where $\chi(S)$ is the Euler characteristic of S.

A smooth map $\phi : (M, J) \rightarrow (M', J')$ from one almost complex manifold to another is said to be (J, J')-**holomorphic** if and only if its derivative $d\phi_x : T_x M \rightarrow T_{\phi(x)} M'$ is complex linear, that is

$$d\phi_x \circ J_x = J'_{\phi(x)} \circ d\phi_x.$$

These are the Cauchy-Riemann equations, and, when (M, J) and (M', J') are both subsets of complex n-space $\mathbb{C}^n$, they are satisfied exactly by the holomorphic maps. An almost complex structure J is said to be **integrable** if it arises from an underlying complex structure on M. This is equivalent to saying that one can choose an atlas for M whose coordinate charts are (J, i)-holomorphic where i is the standard complex structure on $\mathbb{C}^n$. In this case the coordinate changes are holomorphic maps (in the usual sense) between open sets in $\mathbb{C}^n$. When M has dimension 2 a fundamental theorem says that all almost complex structures J on M are integrable. However this is far from true in higher dimensions.

The basic example of an almost complex symplectic manifold is standard Euclidean space $(\mathbb{R}^{2n}, \omega_0)$ with its standard almost complex structure J_0 obtained from the usual identification with $\mathbb{C}^n$. Thus, one sets

$$z_j = x_{2j-1} + ix_{2j}$$

for $j = 1, \ldots, n$ and defines J_0 by

$$J_0(\partial_{2j-1}) = \partial_{2j}, \quad J_0(\partial_{2j}) = -\partial_{2j-1}$$

where $\partial_j = \partial/\partial x_j$ is the standard basis of $T_x\mathbb{R}^{2n}$. Kähler manifolds give another basic example.

1.2 *J*-holomorphic curves

A J-holomorphic curve is a (j, J)-holomorphic map

$$u : \Sigma \to M$$

from a Riemann surface (Σ, j) to an almost complex manifold (M, J).[1] Usually, we will take (Σ, j) to be the Riemann sphere $S^2 = \mathbb{C} \cup \{\infty\}$. In accordance with the terminology of complex geometry it is often convenient to think of the 2-sphere as the complex projective line $\mathbb{C}P^1$. If u is an embedding (that is, an injective immersion) then its image C is a 2-dimensional submanifold of M whose tangent spaces T_xC are J-invariant. Thus each tangent space is a complex line in TM. Conversely, any 2-dimensional submanifold C of M with a J-invariant tangent bundle TC has a J-holomorphic parametrization u. This follows immediately from the fact that the restriction of J to C is integrable.

Note that according to this definition, a curve u is always parametrized. One should contrast this with the situation in complex geometry, where one often defines a curve by requiring it to be the common zero set of a certain number of holomorphic polynomials. Such an approach makes no sense in the non-integrable, almost complex context, since when J is non-integrable there usually are no holomorphic functions $(M, J) \to \mathbb{C}$.

By the taming condition, ω restricts to a positive form on each such line. Therefore C is a symplectic submanifold of M.[2] Conversely, given a 2-dimensional symplectic submanifold of M, it is not hard to construct an ω-tame J such that TC

---

[1] A Riemann surface is a 1-dimensional complex manifold.

[2] A submanifold X of M is said to be symplectic if ω restricts to a non-degenerate form on X.

is J-invariant. (First define J on TC, then extend to the tangent spaces $T_x M$ for $x \in C$, and finally extend the section to the rest of M.) Thus J-holomorphic curves are essentially the same as 2-dimensional symplectic submanifolds of M.

1.3 Moduli spaces

The crucial fact about J-holomorphic curves is that when J is generic they occur in finite dimensional families. These families make up finite dimensional manifolds

$$\mathcal{M}(A, J)$$

which are called **moduli spaces** and whose cobordism classes are independent of the particular J chosen in $\mathcal{J}_\tau(M, \omega)$. Here A is a homology class in $H_2(M, \mathbb{Z})$, and $\mathcal{M}(A, J)$ consists of essentially all J-holomorphic curves $u : \Sigma \to M$ which represent the class A. Although the manifold $\mathcal{M}(A, J)$ is almost never compact, it usually retains enough elements of compactness for one to be able to use it to define invariants.

Chapters 2-6 of this book are taken up with formulating and proving precise versions of the above statements. Here is a brief description of the main results.

Local properties

The first chapter is concerned with local properties of J-holomorphic curves. The key result for future developments is perhaps Proposition 2.3.1, which gives a characterization of those curves which are not multiply-covered. A curve $u : \Sigma \to M$ is said to be **multiply-covered** if it is the composite of a holomorphic branched covering map $(\Sigma, j) \to (\Sigma', j')$ of degree greater than 1 with a J-holomorphic map $\Sigma' \to M$. It is called **simple** if it is not multiply-covered. The multiply-covered curves are often singular points in the moduli space $\mathcal{M}(A, J)$, and so one needs a workable criterion which guarantees that u is simple. We will say that a curve u is **somewhere injective** if there is a point $z \in \Sigma$ such that

$$du(z) \neq 0, \qquad u^{-1}(u(z)) = \{z\}.$$

A point $z \in \Sigma$ with this property is called an **injective point** for u.

Proposition 1.3.1 *Every simple J-holomorphic curve u is somewhere injective. Moreover the set of injective points is open and dense in Σ.*

Fredholm theory

Fix a Riemann surface Σ of genus g and denote by $\mathcal{M}(A, J)$ the set of all *simple* J-holomorphic maps $u : \Sigma \to M$ which represent the class A. In Chapter 3 we prove the following theorem.

Theorem 1.3.2 *There is a subset $\mathcal{J}_{\mathrm{reg}}(A) \subset \mathcal{J}_\tau(M, \omega)$ of the second category (i.e. it contains a countable intersection of open and dense sets) such that for each $J \in \mathcal{J}_{\mathrm{reg}}(A)$ the space $\mathcal{M}(A, J)$ is a smooth manifold of dimension*

$$\dim \mathcal{M}(A, J) = n(2 - 2g) + 2c_1(A).$$

This manifold $\mathcal{M}(A, J)$ carries a natural orientation.

Another important theorem specifies the dependence of $\mathcal{M}(A, J)$ on the choice of J (Theorem 3.1.3).

The basic reason why these theorems are valid is that the Cauchy-Riemann equation

$$du \circ j = J \circ du$$

is elliptic, and hence its linearization is Fredholm. A bounded linear operator L from one Banach space X to another Y is said to be **Fredholm** if it has a finite dimensional kernel and a closed range $L(X)$ of finite codimension in Y. The **index** of L is defined to be the difference in dimension between the kernel and cokernel of L:

$$\mathrm{index}\, L = \dim \ker L - \dim \mathrm{coker} L.$$

An important fact is that the set of Fredholm operators is open with respect to the norm topology and the Fredholm index is constant on each component of the set of Fredholm operators. Thus, it does not change as L varies continuously, though of course the dimension of the kernel and cokernel can change.

Fredholm operators are essentially as well-behaved as finite-dimensional operators and they play an important role in infinite dimensional implicit function theorems. Thus, if $\mathcal{F} : X \to Y$ is a a C^∞-smooth map whose derivative $d\mathcal{F}(x) : X \to Y$ is Fredholm of index k at each point $x \in X$ and if $y \in Y$ is a **regular value** of $\mathcal{F}$ in the sense that $d\mathcal{F}(x)$ is surjective for all $x \in \mathcal{F}^{-1}(y)$, then, just as in the finite-dimensional case, the inverse image

$$\mathcal{F}^{-1}(y)$$

is a smooth manifold of dimension k. An infinite dimensional version of Sard's theorem says that almost all points of Y are regular for $\mathcal{F}$. Technically, they form a set of second category. This theorem remains true if X and Y are Banach manifolds rather than Banach spaces. However it does not extend as stated to other kinds of infinite dimensional vector spaces, such as Fréchet spaces.

The set $\mathcal{J}_{\mathrm{reg}}$ mentioned in the above theorem does consist of the regular values of some Fredholm operator which maps into the space $\mathcal{J}_\tau(M, \omega)$. There are some additional technicalities in the proof which are caused by the fact that $\mathcal{J}_\tau(M, \omega)$ is a Fréchet rather than a Banach manifold. Elements $J \in \mathcal{J}_\tau(M, \omega)$ which belong to the subset $\mathcal{J}_{\mathrm{reg}}$ are often said to be **generic**. An interesting fact is that the taming condition on J is irrelevant here. The above results hold for all almost complex structures J on any compact manifold M.

1.4 Compactness

The next task is to develop an understanding of when the moduli spaces $\mathcal{M}(A, J)$ are compact. Here the taming condition plays an essential role. The symplectic form ω and an ω-tame almost complex structure J together determine a Riemannian metric

$$\langle v, w \rangle = \tfrac{1}{2}(\omega(v, Jw) + \omega(w, Jv))$$

on M and the **energy** of a J-holomorphic curve $u : \Sigma \to M$ with respect to this metric is given by the formula

$$E(u) = \int_\Sigma |du|^2 = \int_\Sigma u^*\omega.$$

Thus the L^2-norm of the derivative of a J-holomorphic curve satisfies a uniform bound which depends only on the homology class A represented by u. This in itself does not guarantee compactness because it is a borderline case for Sobolev estimates. (A uniform bound on the L^p-norms of du with $p > 2$ would guarantee compactness.)

Another manifestation of the failure of compactness can be observed in the fact that the energy $E(u)$ is invariant under conformal rescaling of the metric on Σ. This effect is particularly clear in the case where the domain Σ of our curves is the Riemann sphere $\mathbb{C}P^1$, since here there is a large group of global, rather than local, rescalings. Indeed, the non-compact group $G = \mathrm{PSL}(2, \mathbb{C})$ acts on the Riemann sphere by conformal transformations

$$z \mapsto \frac{az + b}{cz + d}.$$

Thus each element $u \in \mathcal{M}(A, J)$ has a non-compact family of reparametrizations $u \circ \phi$, for $\phi \in G$, and so $\mathcal{M}(A, J)$ itself can never be both compact and non-empty (unless A is the zero class, in which case all the maps u are constant). However, the quotient space $\mathcal{C}(A, J) = \mathcal{M}(A, J)/G$ will sometimes be compact.

Recall that a homology class $B \in H_2(M)$ is called **spherical** if it is in the image of the Hurewicz homomorphism $\pi_2(M) \to H_2(M)$. Here is a statement of the first important theorem in Chapter 4.

Theorem 1.4.1 *Assume that there is no spherical homology class $B \in H_2(M)$ such that $0 < \omega(B) < \omega(A)$. Then the moduli space $\mathcal{M}(A, J)/G$ is compact.*

One proves this by showing that, if u_ν is a sequence in $\mathcal{M}(A, J)$ which has no limit point in $\mathcal{M}(A, J)$, then there is a point $z \in \mathbb{C}P^1$ at which the derivatives $du_\nu(z)$ are unbounded. This implies that, after passage to a subsequence, there is a decreasing sequence U_ν of neighbourhoods of z in $\mathbb{C}P^1$ whose images $u_\nu(U_\nu)$ converge in the limit to a J-holomorphic sphere. If B is the homology class represented by this sphere, then either $\omega(B) = \omega(A)$, in which case the maps u_ν can be reparametrized so that they do converge in $\mathcal{M}(A, J)$, or $\omega(B)$ lies strictly between 0 and $\omega(A)$. This is the process of "bubbling", which occurs in this context in a simple and geometrically clear way.

One consequence of this theorem is that if $\omega(A)$ is already the smallest positive value assumed by ω on spheres then the moduli space $\mathcal{M}(A, J)/G$ is compact. To go further, we need a sharper form of this result, which describes the complete limit of the sequence u_ν. This is Gromov's compactness Theorem 4.4.3, which is stated in Section 4.4. A complete proof is given in Section 4.5.

1.5   Evaluation maps

The invariants which we shall use are built from the evaluation map

$$\mathcal{M}(A, J) \times \mathbb{C}P^1 \to M : (u, z) \mapsto u(z).$$

Note that this factors through the action of the reparametrization group G given by

$$\phi \cdot (u, z) = (u \circ \phi^{-1}, \phi(z)).$$

Hence we get a map defined on the quotient

$$e = e_J : \mathcal{W}(A, J) = \mathcal{M}(A, J) \times_G \mathbb{C}P^1 \to M.$$

Example 1.5.1 Suppose that M is a product $\mathbb{C}P^1 \times V$ with a product symplectic form and let $A = [\mathbb{C}P^1 \times \text{pt}]$. Suppose also that $\pi_2(V) = 0$. Then A generates the group of spherical 2-classes in M, and so $\omega(A)$ is necessarily the smallest value assumed by ω on the spherical classes. Theorems 3.1.2 and 4.3.2 therefore imply that the space $\mathcal{W}(A, J)$ is a compact manifold for generic J. Because $c_1(A) = 2$, in this case the dimension of $\mathcal{W}(A, J)$ is $2n$ and agrees with the dimension of M. Moreover, we will see in Chapter 3 (see Theorem 3.1.3) that different choices of J give rise to cobordant maps e_J. Since cobordant maps have the same degree, this means that the degree of e_J is independent of all choices. Now if $J = i \times J'$ is a product, where i denotes the standard complex structure on $\mathbb{C}P^1$, then it is easy to see that the elements of $\mathcal{M}(A, J)$ have the form

$$u(z) = (\phi(z), v_0)$$

where $v_0 \in V$ and $\phi \in G$. It follows that the map e_J has degree 1 for this choice of J and hence for every J.

Gromov used this fact in [26] to prove his celebrated non-squeezing theorem.

Theorem 1.5.2 *If ψ is a symplectic embedding of the ball $B^{2n}(r)$ of radius r into a cylinder $B^2(\lambda) \times V$, where $\pi_2(V) = 0$, then $r \leq \lambda$.*

Sketch of proof: Embed the disc $B^2(\lambda)$ into a 2-sphere $\mathbb{C}P^1$ of area $\pi\lambda^2 + \varepsilon$, and let ω be the product symplectic structure on $\mathbb{C}P^1 \times V$. Let J' be an ω-tame almost complex structure on $\mathbb{C}P^1 \times V$ which, on the image of ψ, equals the push-forward by ψ of the standard structure J_0 of the ball $B^{2n}(r)$. Since the evaluation map $e_{J'}$ has degree 1, there is a J'-holomorphic curve through every point of $\mathbb{C}P^1 \times V$. In particular, there is such a curve, C' say, through the image $\psi(0)$ of the center of the ball. This curve pulls back by ψ to a J_0-holomorphic curve C through the center of the ball $B^{2n}(r)$. Since J_0 is standard, this curve C is holomorphic in the usual sense and so is a minimal surface in $B^{2n}(r)$. But it is a standard result in the theory of minimal surfaces, that the surface of smallest area which goes through the center of a ball in Euclidean space is the flat disc of area πr^2. Further, because C is holomorphic, it is easy to check that its area is just given by the integral of the standard symplectic form ω_0 over it. Thus

$$\pi r^2 \leq \int_C \omega_0 = \int_{\psi^{-1}(C')} \psi^*(\omega) < \int_{C'} \omega = \omega(A) = \pi\lambda^2 + \varepsilon$$

where the middle inequality holds because $\psi(C)$ is only a part of C'. Since this is true for all $\varepsilon > 0$, the result follows. More details may be found in [26, 43, 36, 52].
□

Often it is useful to evaluate the map u at more than one point and so we shall consider the maps

$$e_p : \mathcal{W}(A, J, p) = \mathcal{M}(A, J) \times_{\mathrm{G}} (\mathbb{C}P^1)^p \to M^p$$

defined by

$$e_p(u, z_1, \ldots, z_p) = (u(z_1), \ldots, u(z_p)).$$

Here p is any positive integer and $M^p = M \times \cdots \times M$. For a generic almost complex structure the domain $\mathcal{W}(A, J, p)$ of this evaluation map is a manifold of dimension

$$\dim \mathcal{W}(A, J, p) = 2n + 2c_1(A) + 2p - 6.$$

In general this manifold will not be compact. However, in many cases one can show that its image $X(A, J, p) = e_p(\mathcal{W}(A, J, p)) \subset M^p$ can be compactified by adding pieces of dimension at least 2 less than that of $\mathcal{W}(A, J, p)$. This is the content of the following theorem. For the purposes of exposition, we state it in a slightly simplified form. In particular, we assume that (M, ω) is **monotone**. This means that there is a positive constant $\lambda > 0$ such that

$$\omega(B) = \lambda c_1(B)$$

for all spherical classes $B \in H_2(M; \mathbb{Z})$, where c_1 is the first Chern class of the complex bundle (TM, J). This hypothesis is stronger than necessary, but has the virtue of being easy to understand.

Theorem 1.5.3 *Let (M, ω) be a monotone compact symplectic manifold and $A \in H_2(M, \mathbb{Z})$.*

(i) *For every $J \in \mathcal{J}_\tau(M, \omega)$ there exists a finite collection of evaluation maps $e_k : \mathcal{W}_k(J) \to M^p$ such that*

$$\bigcap_{\substack{K \subset \mathcal{W}(A, J, p) \\ K \; compact}} \overline{e_p(\mathcal{W}(A, J, p) - K)} \subset \bigcup_k e_k(\mathcal{W}_k(J)).$$

(ii) *There is a set of second category in $\mathcal{J}_\tau(M, \omega)$ such that, for every J in this set, the spaces $\mathcal{W}_k(J)$ are smooth oriented σ-compact manifolds of dimensions*

$$\dim \mathcal{W}_k(J) \leq \dim \mathcal{W}(A, J, p) - 2.$$

A collection of such theorems is stated and discussed in Chapter 5. They are proved in Chapter 6.

1.6   The Gromov-Witten invariants

The above theorem implies that the map $e_p : \mathcal{W}(A, J, p) \to M^p$ represents a well-defined homology class in M^p. Intuitively, its image is an m-chain, where $m = \dim \mathcal{W}(A, J, p)$, whose boundary has dimension at most $m - 2$ and so is not seen from homological point of view. In Chapter 7 this is formalised in the notion of a **pseudo-cycle**. One can show that the homology class represented by the evaluation map e_p does not depend on the choice of J. The **Gromov invariants** are obtained by taking its intersection with cycles of complementary dimension in M^p.

More precisely, let $\alpha = \alpha_1 \times \cdots \times \alpha_p$ be an element of $H_d(M^p)$ where

$$d + \dim \mathcal{W}(A, J, p) = 2np.$$

Then we may choose a representing cycle for α – also denoted by α – which intersects the image of $e_p : \mathcal{W}(A, J, p) \to M^p$ transversely in a finite set of points. We define the Gromov invariant

$$\Phi_A(\alpha_1, \ldots, \alpha_p) = e_p \cdot \alpha$$

to be the number of these intersection points counted with signs according to their orientations. This is simply the number of J-holomorphic curves u in the homology class A which meet each of the cycles $\alpha_1, \ldots, \alpha_p$. If the dimensional condition is not satisfied, we simply set $\Phi_A(\alpha_1, \ldots, \alpha_p) = 0$. Some examples are calculated in Section 7.3. In particular, we sketch Ruan's argument which uses these invariants to show that there are non-deformation equivalent 6-manifolds. The discussion in Sections 7.1 and 7.2 is slightly more technical than the present one mostly because we want to weaken the hypothesis of monotonicity. However, the examples in Section 7.3 should be accessible at this point.

The rest of Chapter 7 is devoted to a discussion of the Gromov-Witten invariant Ψ. Here one fixes a p-tuple of distinct points $z_1, \ldots, z_p$ in $\mathbb{C}P^1$, and counts the number of curves u such that

$$u(z_1) \in \alpha_1, \ldots, u(z_p) \in \alpha_p.$$

Again, the dimensions are chosen so that there will only be finitely many such curves. Because the reparametrization group G is triply transitive, it is possible to do this only if $p \geq 3$. Moreover, in the case $p = 3$ both invariants agree

$$\Phi_A(\alpha_1, \alpha_2, \alpha_3) = \Psi_A(\alpha_1, \alpha_2, \alpha_3).$$

However, when $p > 3$ the two invariants are rather different. In particular, it turns out that the invariants Ψ have much simpler formal properties. For example, we shall see in Section 8.2 that they obey the decomposition rule

$$\Psi_A(\alpha_1, \alpha_2, \alpha_3, \alpha_4) = \sum_B \sum_i \Psi_{A-B}(\alpha_1, \alpha_2, \varepsilon_i) \Psi_B(\phi_i, \alpha_3, \alpha_4),$$

where $B \in H_2(M)$, ε_i runs over a basis for the homology $H_*(M)$, and ϕ_i is the dual basis with respect to the intersection pairing. This turns out to be a crucial ingredient in the proof of associativity of the deformed cup product.

1.7 Quantum cohomology

We show in Chapter 8 how to use the invariants Φ_A to define a quantum deformation of the cup product on the cohomology of a symplectic manifold. To begin with we shall assume that M is monotone. Moreover, we assume that $N \geq 2$ where N is the **minimal Chern number** defined by $\langle[c_1], \pi_2(M)\rangle = N\mathbb{Z}$. If we multiply ω by a suitable constant we may assume that $\lambda = 1/N$ and so $\langle[\omega], \pi_2(M)\rangle = \mathbb{Z}$.

The basic idea in the definition of quantum cohomology is very easy to understand. Let us write $H_*(M)$ and $H^*(M)$ for the integral (co)homology of M modulo torsion. Thus one can think of $H^*(M)$ as the image of $H^*(M, \mathbb{Z})$ in $H^*(M, \mathbb{R})$, and similarly for H^*. The advantage of neglecting torsion is that the group $H^k(M)$, for example, may be identified with the dual $\mathrm{Hom}(H_k(M), \mathbb{Z})$ of $H_k(M)$. Thus a k-dimensional cohomology class a may be described by specifying all the values $\int_\gamma a$ of a on the classes $\gamma \in H_k(M)$.

We define the quantum multiplication $a * b$ of classes $a \in H^k(M)$ and $b \in H^\ell(M)$ as follows. Let $\alpha = \mathrm{PD}(a)$ and $\beta = \mathrm{PD}(b)$ denote their Poincaré duals. Then $a * b$ is defined as the formal sum

$$a * b = \sum_A (a * b)_A q^{c_1(A)/N} \tag{1.1}$$

where q is an auxiliary variable, considered to be of degree $2N$, and the cohomology class $(a * b)_A \in H^{k+\ell-2c_1(A)}(M)$ is defined in terms of the Gromov invariant Φ_A by

$$\int_\gamma (a * b)_A = \Phi_A(\alpha, \beta, \gamma)$$

for $\gamma \in H_{k+\ell-2c_1(A)}(M)$. Note that the classes α, β, γ satisfy the dimension condition

$$2c_1(A) + \deg(\alpha) + \deg(\beta) + \deg(\gamma) = 4n$$

required for the definition of the invariant Φ_A. This shows that $0 \leq c_1(A) \leq 2n$ and hence only finitely many powers of q occur in the formula (1.1). Moreover, since M is monotone, the classes A which contribute to the coefficient of q^d satisfy $\omega(A) = c_1(A)/N = d$, and hence only finitely many can be represented by J-holomorphic curves. This shows that the sum on the right hand side of (1.1) is finite. Since only nonnegative exponents of q occur in the sum (1.1) it follows that $a * b$ is an element of the group

$$\widetilde{QH}^*(M) = H^*(M) \otimes \mathbb{Z}[q],$$

where $\mathbb{Z}[q]$ is the polynomial ring in the variable q of degree $2N$. Extending by linearity, we therefore get a multiplication

$$\widetilde{QH}^*(M) \otimes \widetilde{QH}^*(M) \to \widetilde{QH}^*(M).$$

It turns out that this multiplication is distributive over addition, skew-commutative and associative. The first two properties are obvious, but the last is much more subtle and depends on a gluing argument for J-holomorphic curves.

The quantum cohomology groups $\widetilde{QH}^*(M)$ vanish for $k \leq 0$ and are periodic with period $2N$ for $k \geq 2n$. To get full periodicity, one can consider the groups

$$QH^*(M) = H^*(M) \otimes \mathbb{Z}[q, q^{-1}],$$

where $\mathbb{Z}[q, q^{-1}]$ is the ring of **Laurent polynomials**, which consists of all polynomials in the variables q, q^{-1} with the obvious relation $q \cdot q^{-1} = 1$. With this definition, $QH^k(M)$ is nonzero for both positive and negative k, and there is a natural isomorphism

$$QH^k(M) \to QH^{k+2N}(M),$$

given by multiplication with q, for every $k \in \mathbb{Z}$.

Note that if $A = 0$, then all J-holomorphic curves in the class A are constant. It follows that $\Phi_A(\alpha, \beta, \gamma)$ is just the usual triple intersection $\alpha \cdot \beta \cdot \gamma$. Since $\omega(A) > 0$ for all other A which have J-holomorphic representatives, the constant term in $a * b$ is just the usual cup product.

As an example, let M be complex projective n-space $\mathbb{C}P^n$ with its usual Kähler form. If p is the positive generator of $H^2(\mathbb{C}P^n)$, and if L is the class in H_2 represented by the line $\mathbb{C}P^1$, then the fact that there is a unique line through any two points is reflected in the identity

$$\int_{\mathrm{pt}} (p * p^n)_L = \Phi_L([\mathbb{C}P^{n-1}], \mathrm{pt}, \mathrm{pt}) = 1.$$

Since all the other classes $(p * p^n)_A$ vanish for reasons of dimension, it follows that

$$p * p^n = q$$

and hence the quantum cohomology of $\mathbb{C}P^n$ is given by

$$\widetilde{QH}^*(\mathbb{C}P^n) = \frac{\mathbb{Z}[p, q]}{\langle p^{n+1} = q \rangle}.$$

See Example 8.1.6 for more details. The occurence of the letters p, q is no accident here. In Section 8.3, we describe some recent work of Givental and Kim in which they interpret the quantum cohomology ring of flag manifolds as the ring of functions on a Lagrangian variety.

We end the chapter by explaining how the higher Gromov-Witten invariants are generated by a function S which satisfies the WDVV-equation.

1.8 Novikov rings and Floer homology

If the image Γ of the the Hurewicz homomorphism $\pi_2(M) \to H_2(M)$ has rank bigger than 1 it is possible to modify the definition of quantum cohomology slightly, even in the monotone case. Roughly speaking, one can count J-holomorphic curves which represent different homology classes separately and hence extract more information from the moduli spaces. This leads naturally to tensoring with the Novikov ring Λ_ω associated to the homomorphism $\Gamma \to \mathbb{R}$ induced by ω. This homomorphism is defined by evaluating the symplectic form ω on a homology class $A \in \Gamma \subset H_2(M)$ and hence measures the energy of the J-holomorphic curves which represent A. The Novikov ring is a kind of completion of the group ring which in the case where Γ is 1-dimensional and M is monotone agrees with the ring of Laurent power series in q, q^{-1}. In the monotone case, one can define the deformed cup product using either the polynomial ring $\mathbb{Z}[q]$ or the Novikov ring as coefficients. However, for more general symplectic manifolds, for example in the Calabi-Yau case where $c_1 = 0$, it is necessary to take coefficients in some kind of completed ring, because there may be infinitely many J-holomorphic curves intersecting the cycles α, β, γ, although there will only be finitely many with energy less than any fixed bound. If we also want to carry along the information about the homology class A in order to retain as much information as possible, this leads precisely to the framework of the Novikov ring. All this is explained in much more detail in Chapter 9.

There is a striking similarity of these structures with Floer homology (cf. [18], [20]). In the monotone case, the Floer homology groups of the loop space of M are naturally isomorphic to the the ordinary homology groups of M *rolled up with period* $2N$, which are exactly the quantum cohomology groups in their periodic form, i.e. tensored with $\mathbb{Z}[q, q^{-1}]$. Moreover, in the non-monotone case, the Novikov ring enters into the construction of Floer homology. This was already pointed out by Floer in his original paper [20] and was later rediscovered in a joint paper [32] of the second author with Hofer. There is a natural ring structure on the Floer homology groups given by the *pair-of-pants* construction, and one is led to wonder whether this ring structure agrees with the deformed cup-product. In the last chapter we shall explain the basic ideas in Floer homology, and give an outline of the proof why the ring structure should agree with that of quantum cohomology.

Chapter 2

Local Behaviour

In this chapter we establish the basic properties of J-holomorphic curves. This includes the unique continuation theorem, which asserts that two curves with the same ∞-jet at a point must coincide, and Proposition 2.3.1, which asserts that a simple J-holomorphic curve must have injective points. These are essential ingredients in the transversality theory for J-holomorphic curves discussed in Chapter 3. The other main results are Lemmata 2.2.1 and 2.2.3.

2.1   The generalised Cauchy-Riemann equation

Let (M, J) be an almost complex manifold and (Σ, j) be a Riemann surface. A smooth map $u : \Sigma \to M$ is called J-**holomorphic** if the differential du is a complex linear map with respect to j and J:

$$J \circ du = du \circ j.$$

Sometimes it is convenient to write this equation as $\bar{\partial}_J(u) = 0$. Here, the 1-form

$$\bar{\partial}_J(u) = \frac{1}{2}\left(du + J \circ du \circ j\right) \in \Omega^{0,1}(u^*TM)$$

is the complex antilinear part of du, and takes values in the complex vector bundle $u^*TM = \left\{(z, v) \mid z \in \Sigma,\ v \in T_{u(z)}M\right\}$. Thus there is an infinite dimensional vector bundle

$$\mathcal{E} \to \mathrm{Map}(\Sigma, M)$$

whose fibre at $u \in \mathrm{Map}(\Sigma, M)$ is the space $\mathcal{E}_u = \Omega^{0,1}(u^*TM)$ and $\bar{\partial}_J$ is a section of this bundle. The J-holomorphic curves are the zeros of this section.

As we shall see later, this global form of the equation is useful when one is discussing properties of the moduli space of all J-holomorphic curves. In order to study local properties of these curves, it is useful to write the equation in local coordinates.

By the integrability theorem there exists an open cover $\{U_\alpha\}_\alpha$ of Σ with charts $\alpha : U_\alpha \to \mathbb{C}$ such that $d\alpha(p)jv = id\alpha(p)v$ for $v \in T_p\Sigma$. In particular the transition maps $\alpha \circ \beta^{-1}$ are holomorphic. Such local coordinates are called **conformal**. (Recall that a map ϕ between open subsets of $\mathbb{C}$ is conformal, i.e. preserves angles and orientation, if and only if it is holomorphic.) A smooth map $u : \Sigma \to M$ is J-holomorphic if and only if its local coordinate representations

$$u^\alpha = u \circ \alpha^{-1} : \alpha(U_\alpha) \to M$$

are J-holomorphic with respect to the standard complex structure i on the open set $\alpha(U_\alpha) \subset \mathbb{C}$.

In conformal coordinates $z = s + it$ on Σ the 1-form $\bar{\partial}_J(u)$ is given by

$$\bar{\partial}_J(u^\alpha) = \frac{1}{2}\left(\frac{\partial u^\alpha}{\partial s} + J(u^\alpha)\frac{\partial u^\alpha}{\partial t}\right) ds + \frac{1}{2}\left(\frac{\partial u^\alpha}{\partial t} - J(u^\alpha)\frac{\partial u^\alpha}{\partial s}\right) dt.$$

Hence u is a J-holomorphic curve if and only if in conformal coordinates it satisfies the nonlinear first order partial differential equation

$$\frac{\partial u^\alpha}{\partial s} + J(u^\alpha)\frac{\partial u^\alpha}{\partial t} = 0. \tag{2.1}$$

In the case of the standard complex structure $J = J_0$ on $\mathbb{C}^n = \mathbb{R}^{2n}$ equation (2.1) reduces to the Cauchy-Riemann equations

$$\frac{\partial f}{\partial s} = \frac{\partial g}{\partial t}, \qquad \frac{\partial f}{\partial t} = -\frac{\partial g}{\partial s}.$$

for a smooth map $u = f + ig : \mathbb{C} \to \mathbb{C}^n$. Thus a J_0-holomorphic curve is holomorphic in the usual sense.

Lemma 2.1.1 *Assume Σ is a connected Riemann surface. If two J-holomorphic curves $u, u' : \Sigma \to M$ have the same ∞-jet at a point $z \in \Sigma$ then $u = u'$.*

Proof: Since Σ is connected it suffices to prove this locally. Hence assume that u and u' are solutions of (2.1) on some open neighbourhood Ω of $0 \in \mathbb{C}$. Denote by Δ the standard Laplacian

$$\Delta = \frac{\partial^2}{\partial s^2} + \frac{\partial}{\partial t^2} = (\partial_s)^2 + (\partial_t)^2.$$

Then it is easy to check that u and u' are solutions of the second order quasi-linear equation

$$\Delta u = (\partial_t J(u))\partial_s u - (\partial_s J(u))\partial_t u. \tag{2.2}$$

(Use the fact that $\partial_t(J^2) = (\partial_t J)J + J(\partial_t J) = 0$.) Now the function $v = u' - u$ vanishes to infinite order at $0 \in \Omega$. Because J and its derivatives are bounded it follows that v satisfies differential inequalities of the form

$$|\Delta v(z)| \leq K(|v| + |\partial_s v| + |\partial_t v|),$$

for all $z \in \Omega$. Hence the assertion of the lemma follows from Aronszajn's unique continuation theorem which we now quote.　　　　　　　　　　　　$\square$

Theorem 2.1.2 (Aronszajn) *Let $\Omega \subset \mathbb{C}$ be a connected open set. Suppose that the function $v \in W^{2,2}_{\mathrm{loc}}(\Omega, \mathbb{R}^m)$ satisfies the pointwise estimate*

$$|\Delta v(s,t)| \leq c \left(|v(s,t)| + \left| \frac{\partial v}{\partial s}(s,t) \right| + \left| \frac{\partial v}{\partial t}(s,t) \right| \right)$$

(almost everywhere) and that v vanishes to infinite order at the point $z = 0$ in the sense that

$$\int_{|z| \leq r} |v(z)| = O(r^k)$$

for every $k > 0$. Then $v \equiv 0$.

Here $W^{2,2}_{\mathrm{loc}}(\Omega, \mathbb{R}^m)$ is the Sobolev space of maps whose second derivative is L^2 on each precompact open subset of Ω. Since all the maps which we consider here are smooth, a reader who is unfamiliar with Sobolev spaces can suppose that g is C^∞. This theorem can be viewed as a generalization of the unique continuation theorem for analytic functions. It is proved by Aronszajn in [1] and by Hartman and Wintner in [29].

Remark 2.1.3 The assumptions of Lemma 2.1.1 require that the J-holomorphic curves u and u' are C^∞-smooth. This condition is automatically satisfied when the almost complex structure J is smooth (see Proposition 3.2.2 below). If, however, J is only of class C^ℓ then the J-holomorphic curves will in general also be only of class C^ℓ.[1] In this case there is an analogue of Lemma 2.1.1 which requires that the difference $v = u' - u$ (in local coordinates) vanishes to infinite order at a point z_0 (as in the statement of Aronszajn's theorem). For a proof of this result which does not rely on Aronszajn's theorem we refer to [22]. $\qquad \square$

2.2 Critical points

A **critical point** of a J-holomorphic curve $u : \Sigma \to M$ is a point $z \in \Sigma$ such that $du(z) = 0$. If we think of the image $C = u(\Sigma) \subset M$ as an unparametrized J-holomorphic curve, then a critical point on C is a critical value of u. Points on C which are not critical are called regular or non-singular. In the integrable case critical points of nonconstant holomorphic curves are well known to be isolated. The next lemma asserts this for arbitrary almost complex structures.

Lemma 2.2.1 *Let $u : \Sigma \to M$ be a nonconstant J-holomorphic curve for some compact connected Riemann surface Σ. Then the set*

$$X = u^{-1} \left(\{ u(z) \, | \, z \in \Sigma, \, du(z) = 0 \} \right)$$

of preimages of critical values is finite.

---

[1] In fact, a simple elliptic bootstrapping argument based on equation 2.2 shows that if J is of class C^ℓ with $\ell \geq 1$ then every J-holomorphic curve u of class $W^{1,p}$ with $p > 2$ is necessarily of class $W^{\ell+1,q}$ for every $q < \infty$.

Proof: It suffices to prove that critical points are isolated, and so we may work locally. Thus we may suppose that u is a map of an open neighbourhood Ω of 0 in $\mathbb{C}$ to $\mathbb{C}^n$, which is J-holomorphic for some $J : \mathbb{C}^n \to \mathrm{GL}(2n, \mathbb{R})$, and that

$$u(0) = 0, \qquad du(0) = 0, \qquad u \not\equiv 0, \qquad J(0) = J_0.$$

Write $z = s + it$. Since u is non-constant it follows from Lemma 2.1.1 that the ∞-jet of $u(z)$ at $z = 0$ must be non-zero. Hence there exists an integer $\ell \geq 2$ such that $u(z) = O(|z|^\ell)$ and $u(z) \neq O(|z|^{\ell+1})$. This implies $J(u(z)) = J_0 + O(|z|^\ell)$. Now examine the Taylor expansion of

$$\frac{\partial u}{\partial s} + J(u)\frac{\partial u}{\partial t} = 0$$

up to order $\ell - 1$ to obtain

$$\frac{\partial T_\ell(u)}{\partial s} + J_0\frac{\partial T_\ell(u)}{\partial t} = 0.$$

Here $T_\ell(u)$ denotes the Taylor expansion of u up to order ℓ. It follows that $T_\ell(u) : \mathbb{C} \to \mathbb{C}^n$ is a holomorphic function and there exists a nonzero vector $a \in \mathbb{C}^n$ such that

$$u(z) = az^\ell + O(|z|^{\ell+1}), \qquad \frac{\partial u}{\partial s}(z) = \ell a z^{\ell-1} + O(|z|^\ell).$$

Hence

$$0 < |z| \leq \varepsilon \qquad \Longrightarrow \qquad u(z) \neq 0, \quad du(z) \neq 0$$

with $\varepsilon > 0$ sufficiently small. Hence preimages of critical values of u are isolated. $\square$

We now show how to choose nice coordinates near a regular point of a J-holomorphic curve.

Lemma 2.2.2 *Let $\Omega \subset \mathbb{C}$ be an open neighbourhood of 0 and $u : \Omega \to M$ be a local J-holomorphic curve such that $du(0) \neq 0$. Then there exists a chart $\alpha : U \to \mathbb{C}^n$ defined on a neighbourhood of $u(0)$ such that*

$$\alpha \circ u(z) = (z, 0, \ldots, 0), \qquad d\alpha(u(z))J(u(z)) = J_0 d\alpha(u(z))$$

for $z \in \Omega \cap u^{-1}(U)$.

Proof: Write $z = s + it \in \Omega$ and $w = (w_1, w_2, \ldots, w_n) \in \mathbb{C}^n$ where $w_j = x_j + iy_j$. Shrink Ω if necessary and choose a complex frame of the bundle u^*TM such that

$$Z_1(z), \ldots, Z_n(z) \in T_{u(z)}M, \qquad Z_1 = \frac{\partial u}{\partial s}.$$

Define $\phi : \Omega \times \mathbb{C}^{n-1} \to M$ by

$$\phi(w_1, \ldots, w_n) = \exp_{u(w_1)}\left(\sum_{j=2}^{n} x_j Z_j(w_1) + \sum_{j=2}^{n} y_j J(u(w_1))Z_j(w_1)\right).$$

Then ϕ is a diffeomorphism of a neighbourhood V of zero in $\mathbb{C}^n$ onto a neighbourhood U of $u(0)$ in M. It satisfies $\phi(z_1, 0, \ldots, 0) = u(z_1)$ and

$$\frac{\partial \phi}{\partial x_j} + J(\phi)\frac{\partial \phi}{\partial y_j} = 0$$

at all points $z = (z_1, 0, \ldots, 0)$ and $j = 1, \ldots, n$. Hence the inverse $\alpha = \phi^{-1} : U \to V$ is as required. $\square$

We next start investigating the intersections of two distinct J-holomorphic curves. The most significant results in this connection occur in dimension 4, and are discussed in McDuff [44, 48] for example. For now, we prove a useful result which is valid in all dimensions asserting that intersection points of distinct J-holomorphic curves $u : \Sigma \to M$ and $u' : \Sigma' \to M$ can only accumulate at points which are critical on both curves $C = u(\Sigma)$ and $C' = u'(\Sigma')$. For local J-holomorphic curves this statement can be reformulated as follows.

Lemma 2.2.3 *Let $u, v : \Omega \to M$ be local non-constant J-holomorphic curves defined on an open neighbourhood Ω of $\{0\}$ such that*

$$u(0) = v(0), \qquad du(0) \neq 0.$$

Moreover, assume that there exist sequences $z_\nu, \zeta_\nu \in \Omega$ such that

$$u(z_\nu) = v(\zeta_\nu), \qquad \lim_{\nu \to \infty} z_\nu = \lim_{\nu \to \infty} \zeta_\nu = 0, \qquad \zeta_\nu \neq 0 \neq z_\nu.$$

Then there exists a holomorphic function $\phi : B_\epsilon(0) \to \Omega$ defined in some neighbourhood of zero such that $\phi(0) = 0$ and

$$v = u \circ \phi.$$

Proof: By Lemma 2.2.2 we may assume without loss of generality that $M = \mathbb{C}^n$ and

$$u(z) = (z, 0), \qquad J(w_1, 0) = i$$

where $w = (w_1, \tilde{w})$ with $\tilde{w} \in \mathbb{C}^{n-1}$. Write $v(z) = (v_1(z), \tilde{v}(z))$.

We show first that the ∞-jet of $\tilde{v}$ at $z = 0$ must vanish. Otherwise there would exist an integer $\ell \geq 0$ such that $\tilde{v}(z) = O(|z|^\ell)$ and $\tilde{v}(z) \neq O(|z|^{\ell+1})$. The assumption of the lemma implies $\ell \geq 1$ and hence $J(v(z)) = J_0 + O(|z|^\ell)$. As in the proof of Lemma 2.2.1, consider the Taylor expansion up to order $\ell - 1$ on the left hand side of the equation $\partial_s v + J \partial_t v = 0$ to obtain that $T_\ell(v)$ is holomorphic. Hence

$$v_1(z) = p(z) + O(|z|^{\ell+1}), \qquad \tilde{v}(z) = \tilde{a} z^\ell + O(|z|^{\ell+1})$$

where $p(z)$ is a polynomial of order ℓ and $a \in \mathbb{C}^{n-1}$ is nonzero. This implies that $\tilde{v}(z) \neq 0$ in some neighbourhood of 0 and hence $u(z) \neq v(z)$ in this neighbourhood, in contradiction to the assumption of the lemma. Thus we have proved that the ∞-jet of $\tilde{v}$ at $z = 0$ vanishes.

We prove that $\tilde{v}(z) \equiv 0$. To see this note that, because $J = J_0$ along the axis $\{\tilde{w} = 0\}$,

$$\frac{\partial J(w_1, 0)}{\partial x_1} = \frac{\partial J(w_1, 0)}{\partial y_1} = 0,$$

for all w_1. Hence

$$\left|\frac{\partial J(w)}{\partial x_1}\right| + \left|\frac{\partial J(w)}{\partial y_1}\right| \leq c\,|\tilde{w}|.$$

As in Lemma 2.1.1, it follows easily that

$$|\Delta \tilde{v}| \leq c(|\tilde{v}| + |\partial_s \tilde{v}| + |\partial_t \tilde{v}|).$$

Hence it follows from Aronszajn's theorem that $\tilde{v} \equiv 0$. The required function ϕ is now given by $\phi(z) = v_1(z)$. $\qquad\square$

2.3 Somewhere injective curves

A curve $u : \Sigma \to M$ is said to be **multiply-covered** if it is the composite of a holomorphic branched covering map $(\Sigma, j) \to (\Sigma', j')$ of degree greater than 1 with a J-holomorphic map $\Sigma' \to M$. The curve u is called **simple** if it is not multiply covered. We shall see in the next chapter that the simple J-holomorphic curves in a given homology class form a smooth finite dimensional manifold for generic J. In other words, the multiply covered curves are the exceptional case and they may be singular points in the moduli space of J-holomorphic curves. The proof of this result is based on the observation that every simple J-holomorphic curve is **somewhere injective** in the sense that

$$du(z) \neq 0, \qquad u^{-1}(u(z)) = \{z\}$$

for some $z \in \Sigma$. A point $z \in \Sigma$ with this property is called an **injective point** for u.

Proposition 2.3.1 *Every simple J-holomorphic curve u is somewhere injective. Moreover the set of injective points is open and dense in Σ.*

Proof: Let X be the finite set of critical points of u and

$$X' = u(X)$$

be the corresponding set of critical values. Further, let Q be the set of points of

$$Y = u(\Sigma) - X'$$

where distinct branches of $u(\Sigma)$ meet. By Lemma 2.2.1, Q is a discrete subset of Y (i.e. it has no accumulation points in Y). Thus the set $S = Y - Q$ is an embedded submanifold in M. Let $\iota : S \to M$ denote this embedding. Since only finitely many branches of Y can meet at each point of Q, each such point gives rise to a finite number of ends of S each diffeomorphic to a deleted disc $D - pt$. Therefore, we may add a point to each of these ends and extend ι smoothly over the resulting surface S' to ι'. Because ι' is an immersion, there is a unique complex structure on S' with respect to which ι' is J-holomorphic.

The manifold S' still has ends corresponding to the points in X'. Each such end corresponds to a distinct branch of $\mathrm{Im}\, u$ through a point in X'. Thus it is the conformal image of a deleted disc, and so must have the conformal structure of the deleted disc. Therefore we may form a closed Riemann surface Σ' by adding a point to each end of S'. Further, because u extends over the whole of Σ, the map ι' must extend to a J-holomorphic map $u' : \Sigma' \to M$. This map u' is somewhere injective and u factors as $u' \circ \psi$ where ψ is a holomorphic map $\Sigma \to \Sigma'$. Thus ψ is a branched cover and has degree 1 iff u is somewhere injective.

In the case $\Sigma = \Sigma' = \mathbb{C}P^1$ the above argument can be rephrased in more explicit terms as follows. By Lemma 2.2.2 the set

$$X \subset \mathbb{C}P^1$$

of preimages of critical values of u is finite. Denote by

$$\Gamma_0 \subset (\mathbb{C}P^1 - X) \times (\mathbb{C}P^1 - X)$$

the set of all pairs (z, ζ) such that there exist sequences $z_\nu \to z$ and $\zeta_\nu \to \zeta$ with $u(z_\nu) = u(\zeta_\nu)$ and $(z_\nu, \zeta_\nu) \neq (z, \zeta)$. In other words, Γ_0 is the set of accumulation points of multiple points of u. Isolated self-intersection points are excluded. It follows from Lemma 2.2.3 that Γ_0 is an equivalence relation on $\mathbb{C}P^1 - X$. The projection

$$\pi : \Gamma_0 \to \mathbb{C}P^1 - X$$

onto the first component is a covering and $\pi^{-1}(z)$ is a finite set for every z. This is because $u(\zeta) = u(z)$ for every $\zeta \in \pi^{-1}(z)$ and $u(z)$ is a regular value of u in the sense that $du(\zeta) \neq 0$ for every $\zeta \in u^{-1}(u(z))$. Hence it follows from Lemma 2.2.3 that $u^{-1}(u(z))$ is a finite set. Since $\mathbb{C}P^1 - X$ is connected the number

$$k = \# \left\{ \zeta \in \mathbb{C}P^1 - X \,|\, (z, \zeta) \in \Gamma_0 \right\}$$

is independent of z. Moreover, it follows from Lemma 2.2.3 that each local inverse of π is holomorphic.

If $k = 1$ then it follows from the definition of Γ_0 that the set of noninjective points of u is countable and can accumulate only at the critical set X. In particular u is somewhere injective in this case. Hence assume $k \geq 2$ and extend Γ_0 to its closure

$$\Gamma = \mathrm{cl}(\Gamma_0) \subset \mathbb{C}P^1 \times \mathbb{C}P^1.$$

This set is an equivalence relation on $\mathbb{C}P^1$ and we denote

$$z \sim \zeta \qquad \Longleftrightarrow \qquad (z, \zeta) \in \Gamma.$$

The set X is invariant under this equivalence relation. (If $z \in X$ and $z \sim \zeta$ then $\zeta \in X$.) Hence each point $z \in \mathbb{C}P^1$ carries a natural multiplicity $m(z) \geq 1$ defined as follows. If $z \in \mathbb{C}P^1 - X$ define $m(z) = 1$. If $z \in X$ and $w \in \mathbb{C}P^1 - X$ is sufficiently close to z then all k points in the equivalence class of w are close to X. Define $m(z)$ as the number of points equivalent to w which are close to z. By continuity this number is independent of the choice of w. With this definition we have

$$\sum_{\zeta \sim z} m(\zeta) = k$$

for every $z \in \mathbb{C}P^1$.

We prove that the *holomorphic equivalence relation* Γ gives rise to a meromorphic map $\psi : \mathbb{C}P^1 \to \mathbb{C}P^1$ of degree k such that

$$\psi(z_1) = \psi(z_2) \qquad \Longleftrightarrow \qquad z_1 \sim z_2.$$

To see this define $p : \mathbb{C} \times \mathbb{C} \to \mathbb{C}$ by

$$p(z_1, z_2) = \frac{\prod_{\zeta \sim z_2} (z_1 - \zeta)^{m(\zeta)}}{z_1 - z_2}$$

Then p is a polynomial of degree $k - 1$ in z_1 and an entire function in z_2. Since the entire function $z_2 \mapsto p(z_1, z_2)$ has precisely $k - 1$ zeros it must be a polynomial of degree $k - 1$. Since these zeros agree with those of $z_2 \mapsto p(z_2, z_1)$ both polynomials agree up to a constant factor. Examining the case $z_1 = z_2$ we find

$$p(z_1, z_2) = p(z_2, z_1).$$

Now $p(z_1, z_2) = 0$ iff $z_1 \sim z_2$ and $z_1 \neq z_2$. This implies

$$p(z_1, z_2) = \frac{\psi(z_1) - \psi(z_2)}{z_1 - z_2}$$

for some polynomial $\psi : \mathbb{C} \to \mathbb{C}$ of degree k. To see this choose $\psi(z) = zp(z, 0)$ and examine the zeros on both sides of the above equation. This function ψ is as required. It is a holomorphic map of degree k such that $\psi(z_1) = \psi(z_2)$ implies $u(z_1) = u(z_2)$. Hence there exists a unique map $u' : \mathbb{C}P^1 \to M$ such that

$$u = u' \circ \psi.$$

The map u' is obviously continuous on $\mathbb{C}P^1$ and J-holomorphic on the complement of X. It follows again from the removable singularity theorem for J-holomorphic curves that u' is smooth. $\qquad\qquad\square$

Proposition 2.3.2 *Let $u_j : \mathbb{C}P^1 \to M$ be simple J-holomorphic curves for $j = 1, \ldots, N$ such that $u_i \neq u_j \circ \phi$ for $i \neq j$ and for any fractional linear transformation ϕ. Then there exist points $z_1, \ldots, z_N \in \mathbb{C}P^1$ such that*

$$du_j(z_j) \neq 0, \qquad u_j^{-1}(u_j(z_j)) = \{z_j\}$$

for all j and

$$u_i(z_i) \notin u_j(\mathbb{C}P^1)$$

for $i \neq j$. Moreover, the set of all N-tuples $(z_1, \ldots, z_N) \in (\mathbb{C}P^1)^N$ which satisfy these conditions is open and dense in $(\mathbb{C}P^1)^N$ and its complement has codimension at least 2.

Proof: Since the curves u_j are simple it follows from Proposition 2.3.1 that there exist points $z_j \in \mathbb{C}P^1$ which are injective for u_j. In fact the proof of Proposition 2.3.1 shows that there are at most countably many points z_1 which are not injective for u_1 in the above sense and these can only accumulate only at the finitely many singular points of u_1.

Now we shall prove that for every pair $i \neq j$ the set of N-tuples $(z_1, \ldots, z_N)$ with $u_i(z_i) \neq u_j(\mathbb{C}P^1)$ and $u_j(z_j) \notin u_i(\mathbb{C}P^1)$ is open and dense in $(\mathbb{C}P^1)^N$. To see this denote by $Y_j \subset \mathbb{C}P^1$ the set of points $z \in \mathbb{C}P^1$ such that either $u_j(z)$ is a critical value of u_j or $u_j(z) = u_j(\zeta)$ for some $\zeta \neq z$. In other words, Y_j is the complement of the set of injective points for u_j and hence, by Proposition 2.3.1, it is countable with only finitely many possible accumulation points. Now denote by

$$\Gamma_{ij} \subset (\mathbb{C}P^1 - Y_i) \times (\mathbb{C}P^1 - Y_j)$$

the set of all pairs (z_i, z_j) such that there exist sequences $z_{i\nu} \to z_i$ and $z_{j\nu} \to z_j$ with $u_i(z_{i\nu}) = u_j(z_{j\nu})$ and $(z_{i\nu}, z_{j\nu}) \neq (z_i, z_j)$. In other words, Γ_{ij} is the set of accumulation points of intersections of u_i and u_j. In this case Γ_{ij} is not an equivalence relation but it follows as before from Lemma 2.2.3 that the projections

$$\pi_i : \Gamma_{ij} \to \mathbb{C}P^1 - Y_i, \qquad \pi_j : \Gamma_{ij} \to \mathbb{C}P^1 - Y_j$$

are coverings with locally holomorphic inverses. Moreover, as before the number of points in the fiber $\pi_i^{-1}(z)$ is independent of z and, by definition of Y_i, it is either 0 or 1.

In the case $\pi^{-1}(z) = \emptyset$ it follows from the definition of Γ_{ij} that the set of points $z \in \mathbb{C}P^1$ with $u_i(z) \in u_j(\mathbb{C}P^1)$ is countable with finitely many possible accumulation points and hence its complement is open and dense in $\mathbb{C}P^1$. Since this holds for every pair (i, j) with $i \neq j$ this proves the proposition. Hence assume that the set $\pi_i^{-1}(z)$ consists of precisely one point for each $z \in \mathbb{C}P^1 - Y_i$. Then there is a unique map $\phi : \mathbb{C}P^1 - Y_i \to \mathbb{C}P^1 - Y_j$ such that

$$(z, \phi(z)) \in \Gamma_{ij}$$

for all $z \in \mathbb{C}P^1 - Y_i$ and hence

$$u_i = u_j \circ \phi.$$

Now the map ϕ is holomorphic and, as in the proof of Proposition 2.3.1, it extends to a biholomorphic transformation of $\mathbb{C}P^1$. By assumption such a map does not exist and this proves the proposition. $\square$

It is worth pointing out that all the results of the last two sections remain valid for almost complex structures of class C^ℓ with $\ell \geq 2$. In particular, this holds for the unique continuation theorem, in view of Remark 2.1.3. In the next chapter we shall use the C^ℓ versions of these results to prove the transversality theorems for J-holomorphic curves.

Another important group of results concerns the positivity of intersections of J-holomorphic curves with J-holomorphic submanifolds of codimension 2. This is especially important in the study of symplectic 4-manifolds. These results are not relevant for the theorems proved in this book although they might play an important role in computing quantum cohomology groups or Gromov-invariants in specific cases. We refer the interested reader to [50] and [53].

Chapter 3

Moduli Spaces and Transversality

Let (M, J) be a $2n$-dimensional almost complex manifold without boundary and let Σ be a closed Riemann surface of genus g with complex structure j. In this chapter we examine the space

$$\mathcal{M}(A, J)$$

of all simple J-holomorphic curves which represent a given homology class $A \in H_2(M)$. For simplicity, we will assume that the domain is a fixed Riemann surface (Σ, j). We begin by outlining a proof of the fact that the space $\mathcal{M}(A, J)$ is a finite dimensional manifold for a "generic" set $\mathcal{J}_{\mathrm{reg}}(A)$ of almost complex structures J. For simplicity we shall assume throughout that our symplectic manifold M is compact. However, it is easy to see that all the results of this chapter generalize to the noncompact case.

3.1 The main theorems

Denote by $\mathcal{X} = \mathrm{Map}(\Sigma, M; A)$ the space of all smooth maps $u : \Sigma \to M$ which are somewhere injective and represent the homology class $A \in H_2(M)$. This space can be thought of as a kind of infinite dimensional manifold whose tangent space at $u \in \mathcal{X}$ is the space

$$T_u \mathcal{X} = C^\infty(u^* TM)$$

of all smooth vector fields $\xi(z) \in T_{u(z)} M$ along u. Consider the infinite dimensional vector bundle $\mathcal{E} \to \mathcal{X}$ whose fiber at u is the space

$$\mathcal{E}_u = \Omega^{0,1}(u^* TM)$$

of smooth J-anti-linear 1-forms with values in $u^* TM$. Recall that the complex anti-linear part of du defines a section

$$\bar{\partial}_J : \mathcal{X} \to \mathcal{E}$$

of this vector bundle and the moduli space $\mathcal{M}(A, J) = \bar{\partial}_J^{-1}(0)$ is the intersection with the zero section. In order for $\mathcal{M}(A, J)$ to be a manifold it is required that $\bar{\partial}_J$

be transversal to the zero section. This means that the image of the linearization $d\bar\partial_J(u) : T_u\mathcal{X} \to T_{(u,0)}\mathcal{E}$ of $\bar\partial_J$ is complementary to the tangent space $T_u\mathcal{X}$ of the zero section. Let us write $D\bar\partial_J(u)$ for the composite of $d\bar\partial_J(u)$ with the projection $\pi_u : T_{(u,0)}\mathcal{E} = T_u\mathcal{X} \oplus \mathcal{E}_u \to \mathcal{E}_u$. Then we require that the linearized operator

$$D_u = D\bar\partial_J(u) : C^\infty(u^*TM) \to \Omega^{0,1}(u^*TM)$$

is surjective for every $u \in \mathcal{M}(A, J)$. Formally the operator D_u can be determined by differentiating the local co-ordinate expressions for $\bar\partial_J(u)$ in the direction of a vector field ξ along u. These expressions define a global complex anti-linear 1-form on u^*TM whenever u is a J-holomorphic curve.

It turns out that D_u is an elliptic first order partial differential operators and so is *Fredholm*. This means that D_u has a closed range and finite dimensional kernel and cokernel. The *Fredholm index* of such an operator is defined as the dimension of the kernel minus the dimension of the co-kernel. This index is stable under compact (lower order) perturbations. Now $C^\infty(u^*TM)$ and $\Omega^{0,1}(u^*TM)$ are complex vector spaces but, due to the nonintegrability of J, the operator D_u does not respect the complex structures. However, the complex anti-linear part of the operator D_u is of lower order and removing it does not change the Fredholm index. The complex linear part is a Cauchy-Riemann operator and determines a holomorphic structure on u^*TM. The index of such operators is given by the Riemann-Roch theorem and we obtain

$$\text{index } D_u = n(2 - 2g) + 2c_1(u^*TM).$$

If the operator D_u is onto for every $u \in \mathcal{M}(A, J)$, it follows from the infinite dimensional implicit function theorem that the space $\mathcal{M}(A, J)$ is indeed a finite dimensional manifold whose tangent space at u is the kernel of D_u. We will suppose that the almost complex structure J varies in some space $\mathcal{J}$ of almost complex structures on M which is sufficiently large for the constructions given below to work. For example, $\mathcal{J}$ could be any subset of the space of all smooth structures which is open in the C^∞-topology or, if M has a symplectic form ω, it could be the space of all ω-compatible structures. In all cases $\mathcal{J}$ carries the usual C^∞-topology.

Definition 3.1.1 A point (u, J) is called **regular** if D_u is onto. Given $\mathcal{J}$ as above and $A \in H_2(M, \mathbb{Z})$ we denote by $\mathcal{J}_{\text{reg}}(A)$ the set of all $J \in \mathcal{J}$ such that D_u is onto for every $u \in \mathcal{M}(A, J)$. We also denote by

$$\mathcal{J}_{\text{reg}} = \bigcap_A \mathcal{J}_{\text{reg}}(A)$$

the set of all $J \in \mathcal{J}$ such that D_u is onto for every simple J-holomorphic curve $u : \mathbb{C}P^1 \to M$. We shall use the notation $\mathcal{J}_{\text{reg}} = \mathcal{J}_{\text{reg}}(M, \omega)$ when $\mathcal{J}(M, \omega)$ is the space of ω-compatible almost complex structures. (This is defined in Section 3.2 below).

The first part of the theorem below follows immediately from this definition together with the implicit function theorem proved in Section 3.3. The second part uses a version of Sard's theorem which will be discussed later on.

Theorem 3.1.2 **(i)** *If $J \in \mathcal{J}_{\text{reg}}(A)$ then the space $\mathcal{M}(A, J)$ is a smooth manifold of dimension $n(2 - 2g) + 2c_1(A)$. It carries a natural orientation.*

(ii) *The set $\mathcal{J}_{\mathrm{reg}}(A)$ has second category in $\mathcal{J}$. This means that it contains the intersection of countably many open and dense subsets of $\mathcal{J}$.*

The next task is to discuss the dependence of the manifolds $\mathcal{M}(A, J)$ on the choice of $J \in \mathcal{J}_{\mathrm{reg}}(A)$. A **smooth homotopy** of almost complex structures is a smooth map $[0, 1] \to \mathcal{J} : \lambda \to J_\lambda$. For any such homotopy define

$$\mathcal{M}(A, \{J_\lambda\}_\lambda) = \{(\lambda, u) \mid u \in \mathcal{M}(A, J_\lambda)\}.$$

Given $J_0, J_1 \in \mathcal{J}_{\mathrm{reg}}$ denote by $\mathcal{J}(J_0, J_1)$ the space of all smooth homotopies of almost complex structures connecting J_0 to J_1. In general, even if $\mathcal{J}$ is path-connected, there does not exist a homotopy such that $J_\lambda \in \mathcal{J}_{\mathrm{reg}}$ for every λ. In other words the space $\mathcal{M}(A, J_\lambda)$ may fail to be a manifold for some values of λ. However, there is always a smooth homotopy such that the space $\mathcal{M}(A, \{J_\lambda\}_\lambda)$ is a manifold. Such homotopies are called **regular**. Intuitively speaking, one can think of the space $\mathcal{J}_{\mathrm{reg}}$ of regular almost complex structures as the complement of a subvariety S of codimension 1 in the space $\mathcal{J}$. A smooth homotopy $\lambda \mapsto J_\lambda$ is regular if it is transversal to S. For regular homotopies the space $\mathcal{M}(A, \{J_\lambda\}_\lambda)$ is a manifold with boundary

$$\partial \mathcal{M}(A, \{J_\lambda\}_\lambda) = \mathcal{M}(A, J_1) - \mathcal{M}(A, J_0).$$

The minus sign indicates the reversed orientation.

Theorem 3.1.3 *Assume that $\mathcal{J}$ is path-connected and consider $J_0, J_1 \in \mathcal{J}_{\mathrm{reg}}(A)$. Then there exists a dense set of homotopies*

$$\mathcal{J}_{\mathrm{reg}}(A, J_0, J_1) \subset \mathcal{J}(J_0, J_1)$$

such that, for every $\{J_\lambda\}_\lambda \in \mathcal{J}_{\mathrm{reg}}(A, J_0, J_1)$, the space $\mathcal{M}(A, \{J_\lambda\}_\lambda)$ is a smooth manifold of dimension $n(2 - 2g) + 2c_1(A) + 1$. This manifold carries a natural orientation.

Thus the two moduli spaces $\mathcal{M}(A, J_1)$ and $\mathcal{M}(A, J_0)$ are oriented cobordant. Note, however, that until we establish some version of compactness this does not mean very much. The problem of compactness will be addressed in Chapter 4. That chapter is more or less independent of the present one. However, in order to understand our notation and to have some perspective on Sobolev spaces, we recommend that the reader continue at least until the end of the next section.

3.2 Elliptic regularity

In preparation for the proofs of Theorems 3.1.2 and 3.1.3 we shall first introduce the appropriate Sobolev space framework and discuss (without proof) the relevant elliptic regularity theorems. We shall then prove the first part of Theorem 3.1.2 under the assumption that the operator D_u is surjective for every $u \in \mathcal{M}(A, J)$. In the final and main part of this chapter we shall then address the question of transversality.

We shall work with the space $\mathcal{J} = \mathcal{J}(M, \omega)$ of all almost complex structures on M which are **compatible** with some symplectic form ω on M. This means that

(i) $\omega(v, Jv) > 0$ for all non-zero tangent vectors $v \in TM$, and

(ii) $\omega(Jv, Jw) = \omega(v, w)$ for all $v, w \in TM$.

An equivalent condition is that the formula

$$\langle v, w \rangle = \omega(v, Jw) \tag{3.1}$$

defines a Riemannian metric on M. In fact there is a unique Hermitian form on TM whose real part is given by (3.1) and whose imaginary part is the symplectic form ω. Conversely, any such Hermitian structure on M with imaginary part ω determines a compatible almost complex structure. It is well known that the space $\mathcal{J}(M, \omega)$ is non-empty and contractible. (See, for example, [52].)

Remark 3.2.1 We could equally well work with the larger space $\mathcal{J}_\tau(M, \omega)$ of ω-tame J, consisting of all almost complex structures which satisfy condition (i) above. The proof for this case is slightly simpler than in the ω-compatible case. When it comes to the question of compactness, we shall work with tame J since now the result for tame J implies that for compatible J. In later chapters it will sometimes be convenient to blur the distinction between them and write $\mathcal{J}(M, \omega)$ to denote either space.

Later on it will be important to consider almost complex structures of class C^ℓ, rather than smooth ones, in order to obtain a parameter space with a Banach manifold structure. Hence we shall consider the space

$$\mathcal{J}^\ell = \mathcal{J}^\ell(M, \omega)$$

of all almost complex structures of class C^ℓ which are compatible with ω. We will always assume $\ell \geq 1$ so that in order for the terms in the equation (2.1) to be well-defined functions. Moreover, we shall denote by

$$\mathcal{X}^{k,p}$$

the space of maps $\Sigma \to M$ whose k-th derivatives are of class L^p and which represent the class $A \in H_2(M, \mathbb{Z})$. As explained in more detail in Appendix B, this is the completion of $\mathcal{X}$ with respect to the Sobolev $W^{k,p}$-norm given by the sum of the L^p-norms of all derivatives of u up to order k. These norms can be defined in terms of covariant derivatives using Riemannian metrics on M and Σ. Since all our manifolds are compact this norm does not depend on the choice of these metrics.

In order for the space $\mathcal{X}^{k,p}$ to be well defined we must assume that

$$kp > 2.$$

There are various reasons for this. Firstly, the very definition of $\mathcal{X}^{k,p}$, in terms of local coordinate representations of class $W^{k,p}$, requires this assumption. The point is this: the $W^{k,p}$-norm is well-defined for maps between open sets in Euclidean space, but for a general manifold one needs a space which is invariant under composition with coordinate charts. Now the composition of a $W^{k,p}$-map $u : \mathbb{R}^2 \to \mathbb{R}^{2n}$ with a C^k-diffeomorphism in the source and a C^k-map in the target is again of class $W^{k,p}$ precisely when $kp > 2$. Secondly, the Sobolev embedding theorem asserts that under this condition the space $\mathcal{X}^{k,p}$ embeds into the space of continuous maps from $\Sigma \to M$ and, by Rellich's theorem, this embedding is compact. Thirdly, the condition $kp > 2$ is required to obtain that the product of two $W^{k,p}$-functions is again of this class. In other words, the condition $kp > 2$ is needed to deal with the nonlinearities.

The first key observation is that every J-holomorphic curve of class $W^{1,p}$ with $p > 2$ in a smooth manifold is necessarily smooth. More precisely, we have the following regularity theorem which can be proved by elliptic bootstrapping methods. The details are carried out in Appendix B.

Proposition 3.2.2 (Elliptic regularity) *Assume $J \in \mathcal{J}^{\ell}$ is an almost complex structure of class C^{ℓ} with $\ell \geq 1$. If $u : \Sigma \to M$ is a J-holomorphic curve of class $W^{1,p}$ with $p > 2$ then u is of class $W^{\ell+1,p}$. Moreover, there is a constant $c = c(J, \ell)$ such that*

$$\|u\|_{W^{\ell+1,p}} \leq c\|u\|_{W^{1,p}}.$$

In particular, u is of class C^{ℓ}, and if J is smooth (C^{∞}) then so is u.

This shows that for $J \in \mathcal{J}^{\ell}$ the moduli space of J-holomorphic curves of class $W^{k,p}$ is independent of the choice of k so long as $k \leq \ell + 1$. In fact this condition $k \leq \ell + 1$ is needed for the operator D_u to be well defined on the appropriate Sobolev spaces.

Remark 3.2.3 An important consequence of elliptic regularity is the fact that the kernel and cokernel of an elliptic operator do not depend on the precise choice of the space on which the operator is defined. In our context assume that J is of class C^{ℓ} and u is a J-holomorphic curve and so is of class $W^{\ell+1,q}$ for any q. Now consider the operator

$$D_u : W^{k,p}(u^*TM) \to W^{k-1,p}(\Lambda^{0,1}\Sigma \otimes_J u^*TM)$$

with $k \leq \ell + 1$.[1] Then every element in the kernel of D_u is necessarily of class $W^{\ell+1,q}$ for any q and so the kernel of D_u does not depend on the choice of k and p as long as $k \leq \ell + 1$. Similar remarks apply to the cokernel. In particular, the operator D_u is onto for some choice of k and p if and only if it is onto for all such choices. $\qquad\square$

3.3 Implicit function theorem

In this section we shall formulate an appropriate version of the implicit function theorem. Roughly speaking, this theorem asserts that if u is an *approximate J-holomorphic curve*, in the sense that $\bar{\partial}_J(u)$ is sufficiently small in the L^p-norm, and if the operator D_u is surjective with a uniformly bounded right inverse then

---

[1] Given a vector bundle $E \to M$, we write $C^{\infty}(E)$ for the space of C^{∞}-smooth sections and $W^{k,p}(E)$ for its completion with respect to the $W^{1,p}$-norm, i.e. the k-th derivatives of $W^{k,p}$-sections are of class L^p. Further $\Lambda^{0,1}\Sigma$ denotes the bundle of 1-forms on Σ of type $0, 1$. Hence

$$C^{\infty}(\Lambda^{0,1}\Sigma \otimes_J u^*TM) = \Omega^{0,1}(u^*TM).$$

there exists an actual J-holomorphic curve near u. This theorem will also play an important role in the gluing construction in Appendix A and hence we shall explain it carefully. This section may be skipped at a first reading. The next section may be read independently. We shall assume throughout this section that J is a fixed smooth almost complex structure.

In view of Proposition 3.2.2 it suffices to work with the Sobolev space $\mathcal{X}^{1,p}$ of $W^{1,p}$-maps $u : \Sigma \to M$ for some fixed number $p > 2$. Consider the infinite dimensional Banach space bundle $\mathcal{E}^p \to \mathcal{X}^{1,p}$ whose fiber at $u \in \mathcal{X}^{1,p}$ is the space

$$\mathcal{E}_u^p = L^p(\Lambda^{0,1}T^*\Sigma \otimes_J u^*TM)$$

of complex anti-linear 1-forms on Σ of class L^p which take values in the pullback tangent bundle u^*TM. The nonlinear Cauchy-Riemann equations determine a section

$$\bar{\partial}_J : \mathcal{X}^{1,p} \to \mathcal{E}^p$$

of this bundle whose differential at a J-holomorphic curve $u : \Sigma \to M$ is the operator

$$D_u : W^{1,p}(u^*TM) \to L^p(\Lambda^{0,1}T^*\Sigma \otimes_J u^*TM).$$

This operator is uniquely determined for each J-holomorphic curve u, however its definition for general u depends on a choice of connection on TM. If we take the Levi-Civita connection ∇ of the metric (3.1) which is induced by J then an explicit formula for D_u is

$$D_u\xi = \tfrac{1}{2}(\nabla\xi + J(u)\nabla\xi \circ j) + \tfrac{1}{2}(\nabla_\xi J)(u)\partial_J(u) \circ j. \tag{3.2}$$

An alternative formula in terms of a Hermitian connection and the Nijenhuis tensor is given below. The two formulae only agree for J-holomorphic curves u and will be different for other maps.

Remark 3.3.1 (Formula for D_u) Here is another formula for D_u for a curve u which is J-holomorphic. We will write it in terms of a Hermitian connection ∇ on M, so that parallel translation commutes with J. When J is not integrable, we cannot assume that ∇ is torsion free. However, as is shown in [34], we may assume that it equals $\tfrac{1}{4}N_J$, where N_J is the Nijenhuis torsion tensor which measures the non-integrability of J. In this notation, the operator

$$D_u : W^{k,p}(u^*TM) \to W^{k-1,p}(\Lambda^{0,1}T^*\Sigma \otimes_J u^*TM)$$

is given by the following formula

$$D_u\xi = \tfrac{1}{2}(\nabla\xi + J(u)\nabla\xi \circ j) + \tfrac{1}{8}N_J(\partial_J(u),\xi).$$

A geometric derivation of this formula may be found in McDuff [42], Proposition 4.1, for example. $\square$

The ellipticity of the operator D_u can be expressed in terms of the estimate

$$\|\xi\|_{W^{1,p}} \le c_0 \left(\|D_u\xi\|_{L^p} + \|\xi\|_{L^p}\right) \tag{3.3}$$

This estimate follows from the Calderon-Zygmund inequality (the L^p-estimate for Laplace's operator) which is proved in Appendix B.

Remark 3.3.2 The constant c_0 in (3.3) depends on the choice of a metric on the Riemann surface Σ. In Appendix A we shall see that it is sometimes important to choose a nonstandard metric with part of the volume concentrated in an arbitrarily small ball. But the volume of Σ with respect to all these metrics will be uniformly bounded. Moreover, these metrics are chosen such that the L^p norm of the 1-form $du \in \Omega^1(u^*TM)$ satisfies a uniform bound. $\qquad\square$

The estimate (3.3) implies that the operator D_u has a closed range and a finite dimensional kernel. Now consider the formal adjoint operator

$$D_u^* : W^{1,q}(\Lambda^{0,1}T^*\Sigma \otimes_J u^*TM) \to L^q(u^*TM).$$

where $1/p + 1/q = 1$. This operator is defined by the identity

$$\langle \eta, D_u\xi \rangle = \langle D_u^*\eta, \xi \rangle$$

for $\xi \in C^\infty(u^*TM)$ and $\eta \in \Omega^{0,1}(u^*TM)$. Integration by parts shows that D_u^* is again a first order elliptic operator (with coefficients in the same class as u) and hence satisfies an estimate (3.3) with p replaced by q. Hence the operator D_u^* also has a finite dimensional kernel and a closed range. Moreover, it follows from elliptic regularity that every $\eta \in L^q(\Lambda^{0,1}T^*\Sigma \otimes_J u^*TM)$ which annihilates the range of D_u (i.e. $\langle \eta, D_u\xi \rangle = 0$ for all $\xi \in W^{1,p}(u^*TM)$) is in fact of class $W^{1,q}$ and in the kernel of D_u^*. (See Exercise B.3.5 in Appendix B.) This shows that the operator D_u has a finite dimensional cokernel.

Remark 3.3.3 (D_u^* in local coordinates) Choose conformal coordinates $z = s + it$ on Σ and assume that the Riemannian metric on Σ is in these coordinates given by $\theta^{-2}(ds^2 + dt^2)$ for a smooth function $\theta = \theta(z)$. For example the Fubini-Study metric on $\Sigma = \mathbb{C}P^1$ is in the standard coordinates given by $\theta(z) = 1 + |z|^2$. In such coordinates a complex anti-linear 1-form $\eta \in \Omega^{0,1}(u^*TM)$ can be written in the form

$$\eta = \zeta\, ds - J\zeta\, dt$$

where $\zeta(z) \in T_{u(z)}M$. Recall from (3.2) that the operator D_u is given by the formula

$$D_u\xi = \zeta\, ds - J\zeta\, dt, \qquad \zeta = \tfrac{1}{2}\big(\nabla_s\xi + J\nabla_t\xi + \tfrac{1}{2}(\nabla_\xi J)(\partial_t u + J\partial_s u)\big)$$

where ∇ denotes the Levi-Civita connection of the J-induced metric. A simple calculation shows that the formal adjoint operator is given by

$$D_u^*\eta = \theta^2\big(-\nabla_s\zeta + J\nabla_t\zeta + \tfrac{1}{2}(\nabla_\zeta J)(\partial_t u + J\partial_s u) + \tfrac{1}{2}(\nabla_{\partial_t u - J\partial_s u}J)\zeta\big). \qquad\square$$

We shall now turn to the question of surjectivity of the operator D_u. In view of the preceding discussion this is equivalent to the injectivity of the adjoint operator D_u^*. For the implicit function theorem it is important to have a quantitative expression of surjectivity. Roughly speaking, this means that the norm of a suitable right inverse does not get too large. One possibility for constructing such a right inverse is to take the operator

$$Q_u = D_u^*(D_u D_u^*)^{-1} : L^p(\Lambda^{0,1}T^*\Sigma \otimes_J u^*TM) \to W^{1,p}(u^*TM)$$

One can prove that an inequality of the form $\|\eta\|_{W^{1,q}} \leq c\,\|D_u^*\eta\|_{L^q}$ implies a similar inequality for the operator D_u restricted to the image of D_u^*, a complement of the kernel of D_u. (See for example Lemma 4.5 in [16].) An alternative technique for constructing a right inverse is to reduce the domain $W^{1,p}(u^*TM)$ of D_u, by imposing pointwise conditions on ξ, so that the resulting opoerator is bijective, and then taking Q_u to be the inverse of this restricted operator.

The next theorem is a refined version of the implicit function theorem discussed in the beginning of this section. We actually prove that if u is an *approximate* J-holomorphic curve with *sufficiently surjective* operator D_u, then there are J-holomorphic curves near u and they can be modelled on a neighbourhood of zero in the kernel of D_u. More explicitly, for every sufficiently small $\xi \in \ker D_u$ we can find a unique J-holomorphic curve of the form $v_\xi = \exp_u(\xi + Q_u\eta)$.

Theorem 3.3.4 *Let $p > 2$ and $1/p + 1/q = 1$. Then for every constant $c_0 > 0$ there exist constants $\delta > 0$ and $c > 0$ such that the following holds. Let $u : \Sigma \to M$ be a $W^{1,p}$-map and $Q_u : L^p(\Lambda^{0,1}T^*\Sigma \otimes_J u^*TM) \to W^{1,p}(u^*TM)$ be a right inverse of D_u such that*

$$\|Q_u\| \leq c_0, \quad \|du\|_{L^p} \leq c_0, \quad \left\|\bar\partial_J(u)\right\|_{L^p} \leq \delta$$

with respect to a metric on Σ such that $\mathrm{Vol}(\Sigma) \leq c_0$. Then for every $\xi \in \ker D_u$ with $\|\xi\|_{L^p} \leq \delta$ there exists a section $\widehat\xi = Q_u\eta \in \mathrm{W}^{1,p}(u^*TM)$ such that

$$\bar\partial_J(\exp_u(\xi + Q_u\eta)) = 0, \qquad \|Q_u\eta\|_{\mathrm{W}^{1,p}} \leq c\left\|\bar\partial_J(\exp_u(\xi))\right\|_{\mathrm{L}^p}.$$

Proof: The proof is an application of the implicit function theorem for the map

$$\mathcal{F} : W^{1,p}(u^*TM) \to L^p(\Lambda^{0,1}T^*\Sigma \otimes_J u^*TM)$$

defined by

$$\mathcal{F}(\xi) = \Phi_\xi(\bar\partial_J(\exp_u(\xi))) \tag{3.4}$$

where

$$\Phi_\xi : \mathcal{E}^p_{\exp_u(\xi)} \to \mathcal{E}^p_u$$

denotes parallel transport along the geodesic $\tau \mapsto \exp_u(\tau\xi)$. The map $\mathcal{F}$ is smooth and its derivatives are controlled by the L^p-norm of du. The differential at zero is given by

$$d\mathcal{F}(0) = D_u$$

and our assumptions guarantee that this operator is onto and has a right inverse $Q_u : L^p(\Lambda^{0,1}T^*\Sigma \otimes_J u^*TM) \to W^{1,p}(u^*TM)$ such that

$$D_uQ_u = \mathbb{1}, \qquad \|Q_u\| \leq c_0.$$

Moreover the function $\mathcal{F}$ satisfies a quadratic estimate of the form

$$\left\|\mathcal{F}(\xi + \widehat\xi) - \mathcal{F}(\xi) - d\mathcal{F}(\xi)\widehat\xi\right\|_{L^p} \leq c_1 \left\|\widehat\xi\right\|_{L^\infty} \left\|\widehat\xi\right\|_{W^{1,p}}. \tag{3.5}$$

Putting these estimates together we can find the required solution $\widehat\xi = Q_u\eta$ of

$$\mathcal{F}(\xi + Q_u\eta) = 0\ .$$

by the contraction mapping principle for the contraction $\eta \mapsto \eta - \mathcal{F}(\xi + Q_u\eta)$. The solution is the limit of the sequence η_ν generated by the standard iteration

$$\eta_{\nu+1} = \eta_\nu - \mathcal{F}(\xi + Q_u\eta_\nu)$$

with $\eta_0 = 0$. With $\xi_\nu = \xi + Q_u\eta_\nu$ this can also be interpreted as the Newton type iteration $\xi_{\nu+1} = \xi_\nu - Q_u\mathcal{F}(\xi_\nu)$. Details of the convergence proof are standard and are left to the reader. $\qquad\square$

We next show that the J-holomorphic curve which we constructed in the previous theorem from a given ξ is actually unique within a C^0-neighbourhood of u. In the following proposition the map u again plays the role of an approximate J-holomorphic curve, and one should think of $v_0 = \exp_u(\xi + \widehat{\xi})$ as the solution constructed from ξ.

Proposition 3.3.5 *Let $p > 2$ and $1/p + 1/q = 1$. Then for every constant $c_0 > 0$ there exists a constant $\delta > 0$ such that the following holds. Let $u : \Sigma \to M$ be a $W^{1,p}$-map and $Q_u : L^p(\Lambda^{0,1}T^*\Sigma \otimes_J u^*TM) \to W^{1,p}(u^*TM)$ be a right inverse of D_u such that*

$$\|Q_u\| \le c_0, \qquad \|du\|_{L^p} \le c_0$$

with respect to a metric on Σ such that $\mathrm{Vol}(\Sigma) \le c_0$. If $v_0 = \exp_u(\xi_0)$ and $v_1 = \exp_u(\xi_1)$ are J-holomorphic curves such that $\xi_0, \xi_1 \in W^{1,p}(u^*TM)$ satisfy

$$\|\xi_0\|_{W^{1,p}} \le \delta, \qquad \|\xi_1\|_{W^{1,p}} \le c_0,$$

and

$$\|\xi_1 - \xi_0\|_{L^\infty} \le \delta, \qquad \xi_1 - \xi_0 \in \mathrm{im}\, Q_u,$$

then $v_0 = v_1$.

Proof: Choose $\eta = D_u\widehat{\xi} \in L^p(\Lambda^{0,1}T^*\Sigma \otimes_J u^*TM)$ so that

$$\widehat{\xi} = \xi_1 - \xi_0 = Q_u\eta$$

and note that $D_u\eta = \widehat{\xi}$. Let $\xi \mapsto \mathcal{F}(\xi)$ be the map defined in the proof of Theorem 3.3.4. Then $\mathcal{F}(\xi_0) = \mathcal{F}(\xi_1) = 0$. We use the quadratic estimate (3.5) to obtain

$$
\begin{aligned}
\left\|\widehat{\xi}\right\|_{W^{1,p}} &= \left\|Q_u\eta\right\|_{W^{1,p}} \\
&\le c_0 \left\|\eta\right\|_{L^p} \\
&= c_0 \left\|D_u\widehat{\xi}\right\|_{L^p} \\
&= c_0 \left\|d\mathcal{F}(0)\widehat{\xi}\right\|_{L^p} \\
&= c_0 \left\|\mathcal{F}(\xi_1) - \mathcal{F}(\xi_0) - d\mathcal{F}(0)\widehat{\xi}\right\|_{L^p} \\
&\le c_0 \left\|\mathcal{F}(\xi_0 + \widehat{\xi}) - \mathcal{F}(\xi_0) - d\mathcal{F}(\xi_0)\widehat{\xi}\right\|_{L^p} \\
&\quad + c_0 \left\|(d\mathcal{F}(\xi_0) - d\mathcal{F}(0))\widehat{\xi}\right\|_{L^p} \\
&\le c_0 c_1 \left\|\widehat{\xi}\right\|_{L^\infty} \left\|\widehat{\xi}\right\|_{W^{1,p}} + c_2\delta \left\|\widehat{\xi}\right\|_{W^{1,p}} \\
&\le c_3\delta \left\|\widehat{\xi}\right\|_{W^{1,p}}.
\end{aligned}
$$

If $c_3\delta < 1$ then $\xi_1 - \xi_0 = \widehat{\xi} = 0$ and this proves the proposition. $\square$

Proof of Theorem 3.1.2 (i):   If $u \in \mathcal{M}(A, J)$ and and $J \in \mathcal{J}_{\mathrm{reg}}(A)$ then, by assumption, the Fredholm operator D_u is surjective. Hence u satisfies the requirements of Theorem 3.3.4 and hence there exists a unique map

$$f : \ker D_u \to \operatorname{im} D_u^*$$

such that

$$\bar{\partial}_J(\exp_u(\xi + f(\xi))) = 0$$

for every sufficiently small section $\xi \in \ker D_u$. Moreover, the map f is smooth and, by Proposition 3.3.5, every J-holomorphic curve v, which is bounded in $W^{1,p}$ and sufficiently close to u in the L^∞-norm, is of the form

$$v = \exp_u(\xi + f(\xi))$$

with $\xi \in \ker D_u$. To see this write $v = \exp_u(\zeta)$ where $\zeta \in W^{1,p}(u^*TM)$ with $\|\zeta\|_{W^{1,p}} \le c_0$. Now write $\zeta = \xi + D_u^*\eta$ with $D_u\xi = 0$. Then, by Lemma 4.3.1, the $W^{1,p}$-norm of ξ is controlled by its L^p-norm and is therefore small. Hence we can apply Proposition 3.3.5 to obtain $D_u^*\eta = f(\xi)$.

Thus we have proved that the map

$$\ker D_u \to \mathcal{M}(A, J) : \xi \mapsto \exp_u(\xi + f(\xi))$$

defines a local coordinate chart of the Moduli space $\mathcal{M}(A, J)$. We leave it to the reader to check that the transition functions obtained from two nearby elements u and v are smooth. This proves that the moduli space $\mathcal{M}(A, J)$ is a smooth manifold whose dimension agrees with the dimension of the kernel of D_u and therefore with the index of D_u which is $n(2 - 2g) + 2c_1(A)$.

Orientations

To understand why the moduli space has a canonical orientation, observe first that the tangent space $T_u\mathcal{M}(A, J)$ is just the kernel of the operator D_u. Now, according to Remark 3.3.1, the operator D_u is the sum of two parts: $\nabla\xi + J(u)\nabla\xi \circ j$ and $\frac{1}{4}N_J(\partial_J(u), \xi)$. The first of these has order 1 and commutes with J, while the second has order 0 and anti-commutes with J. Hence the kernel of D_u will in general not be invariant under J and so J might not determine an almost complex structure on $T_u\mathcal{M}(A, J)$. However, by multiplying the second part of D_u by a constant which tends to 0, one can homotop D_u through Fredholm operators to a Fredholm operator which does commute with J. The resulting operator

$$C^\infty(u^*TM) \to \Omega^{0,1}(u^*TM) : \xi \mapsto \bar{\partial}_{u,\nabla}\xi = \nabla\xi + J(u)\nabla\xi \circ j$$

is precisely the Dolbeault $\bar{\partial}$-operator. It determines a holomorphic structure on the complex vector bundle u^*TM and its index

$$\operatorname{index} \bar{\partial}_{u,\nabla} = \dim H_{\bar{\partial}}^0(\Sigma, u^*TM) - \dim H_{\bar{\partial}}^{0,1}(\Sigma, u^*TM)$$

is given by the Riemann-Roch theorem:

$$\operatorname{index} D_u = \operatorname{index} \bar{\partial}_{u,\nabla} = 2n + 2c_1(A).$$

(See for example page 246 in Griffiths and Harris [25].)

We now return to the question of orientations. The determinant

$$\det(D) = \Lambda^{\max}(\ker D) \otimes \Lambda^{\max}(\ker D^*)$$

of a Fredholm operator $D : X \to Y$ between complex Banach spaces carries a natural orientation whenever the operator D is complex linear. In our case the operator D_u lies in a component of the space of Fredholm operators which contains a complex linear operator and hence its determinant line $\det(D_u) = \Lambda^{\max}(\ker D_u)$ also carries a natural orientation and this determines an orientation of $T_u\mathcal{M}(A, J) = \ker D_u$. Similar arguments were used by Donaldson in [11] for the orientation of Yang-Mills moduli spaces and also by Ruan [64] in the present context. A slightly different line of argument was used by McDuff in [42] where she established the existence of a canonical stable almost complex structure[2] on compact subsets of the moduli space $\mathcal{M}(A, J)$. $\qquad\Box$

Remark 3.3.6 Note that if J is integrable, then D_u commutes with J, and so J induces an (integrable) almost complex structure on $\mathcal{M}(A, J)$. This is, of course, compatible with the orientation described above. $\qquad\Box$

3.4 Transversality

The proofs of Theorems 3.1.2 and 3.1.3 are based on an infinite dimensional version of Sard's theorem which is due to Smale [80]. Roughly speaking, this says that if X and Y are Banach manifolds and $\mathcal{F} : X \to Y$ is a Fredholm map of index k then the set Y_{reg} of regular values of $\mathcal{F}$ is of the second category, provided that $\mathcal{F}$ is sufficently differentiable (it should be at least C^{k+1}). As in the finite dimensional case a point $y \in Y$ is called a regular value if the linearized operator $D\mathcal{F}(x) : T_xX \to T_yY$ is surjective whenever $\mathcal{F}(x) = y$. It then follows from the implicit function theorem that $\mathcal{F}^{-1}(y)$ is a k-dimensional manifold for every $y \in Y_{\mathrm{reg}}$.

Our strategy is now to prove that the set of regular almost complex structures of class C^ℓ is generic with respect to the C^ℓ-topology and then to take the intersection of these sets over all ℓ. This approach is due to Taubes [81]. We shall begin by discussing the **universal moduli space**

$$\mathcal{M}^\ell(A, \mathcal{J}) = \left\{ (u, J) \in \mathcal{X}^{k,p} \times \mathcal{J}^\ell \,\middle|\, \bar{\partial}_J(u) = 0 \right\}$$

of all J-holomorphic curves where J varies over the space $\mathcal{J}^\ell = \mathcal{J}^\ell(M, \omega)$ of all almost complex structures of class C^ℓ which are compatible with ω. Here $\ell \geq 1$ and, as above, $\mathcal{X}^{k,p}$ denotes the space of $W^{k,p}$-maps $u : \Sigma \to M$ with $p > 2$ and $1 \leq k \leq \ell$. Recall from Proposition 3.2.2 that the space $\mathcal{M}^\ell(A, \mathcal{J})$ is independent of the choice of k and p because every J-holomorphic curve is of class C^ℓ whenever J is of class C^ℓ. The parameter space $\mathcal{J}^\ell$ is a smooth Banach manifold. Its tangent

---

[2]A manifold X is said to have a stable almost complex structure if there is a number k such that the Whitney sum $TX \oplus \mathbf{R}^k$ of the tangent bundle TX with the trivial k-dimensional real bundle carries an almost complex structure.

space $T_J \mathcal{J}^\ell$ at J consists of C^ℓ-sections Y of the bundle $\mathrm{End}(TM, J, \omega)$ whose fiber at $p \in M$ is the space of linear maps $Y : T_p M \to T_p M$ such that

$$YJ + JY = 0, \qquad \omega(Yv, w) + \omega(v, Yw) = 0.$$

This space of C^ℓ-sections is a Banach space and gives rise to a local model for the space $\mathcal{J}^\ell$ via $Y \mapsto J \exp(-JY)$. The corresponding space of C^∞-sections is not a Banach space but only a Fréchet space. Our convention is that spaces with no superscripts consist of elements which are C^∞-smooth.

Proposition 3.4.1 *For every class $A \in H_2(M, \mathbb{Z})$ and every integer $\ell \geq 1$ the universal moduli space $\mathcal{M}^\ell(A, \mathcal{J})$ is a smooth Banach manifold.*

Proof: The map $(u, J) \mapsto \bar{\partial}_J(u)$ defines a section of the infinite dimensional vector bundle $\mathcal{E}^{k-1,p} \to \mathcal{X}^{k,p} \times \mathcal{J}^\ell$. Denote this section by

$$\mathcal{F} : \mathcal{X}^{k,p} \times \mathcal{J}^\ell \to \mathcal{E}^{k-1,p}, \qquad \mathcal{F}(u, J) = \bar{\partial}_J(u).$$

The fiber of $\mathcal{E}^{k-1,p}$ at (u, J) is the space

$$\mathcal{E}^{k-1,p}_{(u,J)} = W^{k-1,p}(\Lambda^{0,1}T^*\Sigma \otimes_J u^*TM)$$

of J-anti-linear 1-forms on Σ of class $W^{k-1,p}$ with values in the bundle $u^*TM \to \Sigma$. We must prove that the differential

$$DF(u, J) : W^{k,p}(u^*TM) \times C^\ell(\mathrm{End}(TM, J, \omega))$$
$$\to W^{k-1,p}(\Lambda^{0,1}T^*\Sigma \otimes_J u^*TM)$$

at a zero (u, J) is surjective whenever u is simple. This differential is given by the formula

$$D\mathcal{F}(u, J)(\xi, Y) = D_u \xi + \tfrac{1}{2} Y(u) \circ du \circ j$$

for $\xi \in W^{k,p}(u^*TM)$ and $Y \in C^\ell(\mathrm{End}(TM, J, \omega))$. The exact formula for D_u is not really relevant: all that matters is that it is a first order differential operator which is Fredholm. We do however point out that the formula (3.2) in holomorphic coordinates $z = s + it$ on Σ and local coordinates on M specializes to

$$D_u \xi = \zeta \, ds - J(u)\zeta \, dt, \qquad \zeta = \partial_s \xi + J(u)\partial_t \xi + (\partial_\xi J(u))\partial_t u$$

whenever u is J-holomorphic. Recall from Proposition 3.2.2 that the J-holomorphic curve u is in fact of class $W^{\ell+1,p}$ and hence, since $p > 2$, of class C^ℓ. Hence the coefficients of the first order terms in D_u are of class C^ℓ and those of the zero order terms are of class $C^{\ell-1}$.

Since D_u is Fredholm the operator $D\mathcal{F}(u, J)$ has a closed range and it suffices to prove that this range is dense. We prove this first for $k = 1$. If the range is not dense then, by the Hahn-Banach theorem, there exists a nonzero $\eta \in L^q(\Lambda^{0,1}T^*\Sigma \otimes_J u^*TM)$ with $1/p + 1/q = 1$ which annihilates the range of $D\mathcal{F}(u, J)$. This means that

$$\int_\Sigma \langle \eta, D_u \xi \rangle = 0 \tag{3.6}$$

for every $\xi \in W^{1,p}(u^*TM)$ and

$$\int_\Sigma \langle \eta, Y(u) \circ du \circ j \rangle = 0 \qquad (3.7)$$

for every $Y \in C^\ell(\mathrm{End}(TM, J, \omega))$. The first equation asserts that the η is a weak solution of $D_u^* \eta = 0$. Since the coefficients of the first order terms in D_u are at least C^ℓ, the same is true for the adjoint operator D_u^* and it follows by elliptic regularity that η is of class $W^{\ell+1,r}$ for any $r > 0$ and $D_u^* \eta = 0$. (Here the relevant elliptic regularity statement is that if D is a first order elliptic operator with C^ℓ-coefficients and if $D\eta \in W^{k,p}$ then $\eta \in W^{k+1,p}$ for $k \le \ell$.) Hence

$$0 = D_u D_u^* \eta = \Delta \eta + \text{lower order terms}$$

and it follows from Aronszajn's theorem 2.1.2 that if η vanishes on some open set then $\eta \equiv 0$.

Since u is simple there exists, by Proposition 2.3.1, a point $z_0 \in \Sigma$ such that

$$du(z_0) \ne 0, \qquad u^{-1}(u(z_0)) = \{z_0\}.$$

We shall prove that η vanishes at z_0. Assume otherwise. Then it is easy to see that there exists a $Y_0 \in \mathrm{End}(T_{u(z_0)}M, J_{u(z_0)}, \omega_{u(z_0)})$ such that

$$\langle \eta(z_0), Y_0 \circ du(z_0) \circ j(z_0) \rangle \ne 0.$$

Now use a smooth cutoff function to find a section $Y \in C^\ell(\mathrm{End}(TM, J, \omega))$ supported near $u(z_0)$ such that $Y_{u(z_0)} = Y_0$. For such a section the left hand side of (3.7) does not vanish. This contradiction shows that $\eta(z_0) = 0$. The same argument shows that η vanishes in a neighbourhood of z_0 and hence $\eta \equiv 0$. This shows that $D\mathcal{F}(u, J)$ has a dense range and is therefore onto in the case $k = 1$.[3]

To prove that $D\mathcal{F}(u, J)$ is onto for general k let $\eta \in W^{k-1,p}(\Lambda^{0,1}T^*\Sigma \otimes_J u^*TM)$ and choose, by surjectivity for $k = 1$, a pair $\xi \in W^{1,p}(u^*TM)$ and $Y \in C^\ell(\mathrm{End}(TM, J, \omega))$ such that

$$D\mathcal{F}(u, J)(\xi, Y) = \eta.$$

Then the above formula for $D\mathcal{F}(u, J)$ shows that

$$D_u \xi = \eta - \tfrac{1}{2} Y(u) \circ du \circ j \in W^{k-1,p}$$

and by elliptic regularity $\xi \in W^{k,p}(u^*TM)$. Hence $D\mathcal{F}(u, J)$ is onto for every pair $(u, J) \in \mathcal{M}^\ell(A, \mathcal{J})$. Because D_u is a Fredholm operator it is easy to prove that the operator $D\mathcal{F}(u, J)$ has actually a right inverse. Hence it follows from the infinite dimensional implicit function theorem that the space $\mathcal{M}^\ell(A, \mathcal{J})$ is an infinite dimensional manifold. $\qquad \square$

---

[3]In this argument we did not actually need Aronszajn's theorem since we have proved directly that η vanishes on the open and dense set of all injective points of u. However, in some situations it is useful to keep J fixed on some subset of M and vary it in an open set which intersects the image of every nonconstant J-holomorphic curve. Then our argument only shows that η vanishes on some open set, and so we need Aronszajn's theorem to be able to conclude that it vanishes everywhere.

Proof of Theorem 3.1.2 (ii): Consider the projection

$$\pi : \mathcal{M}^{\ell}(A, \mathcal{J}) \to \mathcal{J}^{\ell}$$

as a map between separable Banach manifolds. The tangent space

$$T_{(u,J)}\mathcal{M}^{\ell}(A, \mathcal{J})$$

consists of all pairs (ξ, Y) such that

$$D_u \xi + \tfrac{1}{2}Y(u) \circ du \circ j = 0.$$

Moreover, the derivative

$$d\pi(u, J) : T_{(u,J)}\mathcal{M}^{\ell}(A, \mathcal{J}) \to T_J \mathcal{J}$$

is just the projection $(\xi, Y) \mapsto Y$. Hence the kernel of $d\pi(u, J)$ is isomorphic to the kernel of D_u. Moreover, its image consists of all Y such that $Y(u) \circ du \circ j \in \operatorname{im} D_u$. This is a closed subspace of $T_J \mathcal{J}$ and, since $D\mathcal{F}(u, J)$ is onto, it has the same (finite) codimension as the image of D_u. It follows that $d\pi(u, J)$ is a Fredholm operator and has the same index as D_u. Moreover the operator $d\pi(u, J)$ is onto precisely when D_u is onto. Hence a regular value J of π is an almost complex structure with the property that D_u is onto for every simple J-holomorphic curve $u \in \mathcal{M}^{\ell}(u, J) = \pi^{-1}(J)$. In other words the set

$$\mathcal{J}^{\ell}_{\mathrm{reg}} = \left\{ J \in \mathcal{J}^{\ell} \mid D_u \text{ is onto for all } u \in \mathcal{M}^{\ell}(u, J) \right\}$$

of regular almost complex structures is precisely the set of regular values of π. By the Sard-Smale theorem, this set is of the second category in the sense of Baire (a countable intersection of open and dense sets). Here we use the fact that the manifold $\mathcal{M}^{\ell}(A, \mathcal{J})$ and the projection π are of class $C^{\ell-1}$. Hence we can apply the Sard-Smale theorem whenever $\ell - 2 \geq \operatorname{index} D_u = \operatorname{index} \pi$. Thus we have proved that the set $\mathcal{J}^{\ell}_{\mathrm{reg}}$ is dense in $\mathcal{J}^{\ell}$ with respect to the C^{ℓ}-topology. We shall now use the following argument, which was suggested to us by Taubes [81], to deduce that $\mathcal{J}_{\mathrm{reg}}$ is of the second category in $\mathcal{J}$ with respect to the C^{∞}-topology.

Consider the set

$$\mathcal{J}_{\mathrm{reg,K}} \subset \mathcal{J}(M, \omega)$$

of all smooth almost complex structures $J \in \mathcal{J}(M, \omega)$ such that the operator D_u is onto for every J-holomorphic curve $u : \mathbb{C}P^1 \to M$ which satisfies

$$\|du\|_{L^{\infty}} \leq K \tag{3.8}$$

and for which there exists a point $z \in \mathbb{C}P^1$ such that

$$\inf_{\zeta \neq z} \frac{d(u(z), u(\zeta))}{d(z, \zeta)} \geq \frac{1}{K}. \tag{3.9}$$

The latter condition guarantees that u is simple. Moreover, every simple J-holomorphic curve $u : \mathbb{C}P^1 \to M$ satisfies these two conditions for some value of $K > 0$ and some point $z \in \mathbb{C}P^1$. Hence

$$\mathcal{J}_{\mathrm{reg}} = \bigcap_{K > 0} \mathcal{J}_{\mathrm{reg,K}}.$$

We shall prove that each set $\mathcal{J}_{\mathrm{reg},K}$ is open and dense in $\mathcal{J}(M,\omega)$ with respect to the C^∞-topology. We first prove that $\mathcal{J}_{\mathrm{reg},K}$ is open or, equivalently, that its complement is closed. Hence assume that the sequence $J_\nu \notin \mathcal{J}_{\mathrm{reg},K}$ converges to $J \in \mathcal{J}(M,\omega)$ in the C^∞-topology. Then there exists, for every ν, a sequence of J_ν-holomorphic curves $u_\nu : \mathbb{C}P^1 \to M$ which satisfy (3.8) and (3.9) for some $z_\nu \in \mathbb{C}P^1$ such that the operator D_{u_ν} is not surjective. It follows from standard elliptic bootstrapping arguments that there exists a subsequence $u_{\nu'}$ which converges, uniformly with all derivatives, to a smooth J-holomorphic curve $u : \mathbb{C}P^1 \to M$ (see Appendix B). Choose the subsequence such that $z_{\nu'}$ converges to z. Then the limit curve u satisfies the conditions (3.8) and (3.9) for this point z and, moreover, since the operators D_{u_ν} are all not surjective, it follows that D_u cannot be surjective either. This shows that $J \notin \mathcal{J}_{\mathrm{reg},K}$ and thus we have proved that the complement of $\mathcal{J}_{\mathrm{reg},K}$ is closed in the C^∞-topology.

Now we prove that the set $\mathcal{J}_{\mathrm{reg},K}$ is dense in $\mathcal{J}(M,\omega)$ with respect to the C^∞-topology. To see this note first that, by Remark 3.2.3,

$$\mathcal{J}_{\mathrm{reg},K} = \mathcal{J}^\ell_{\mathrm{reg},K} \cap \mathcal{J}$$

where $\mathcal{J}^\ell_{\mathrm{reg},K} \subset \mathcal{J}^\ell(M,\omega)$ is the set of all C^ℓ almost complex structures $J \in \mathcal{J}^\ell(M,\omega)$ such that the operator D_u is onto for every J-holomorphic curve $u : \mathbb{C}P^1 \to M$ of class C^ℓ with with $\|du\|_{L^\infty} \leq K$. Now the same argument as above shows that $\mathcal{J}^\ell_{\mathrm{reg},K}$ is open in $\mathcal{J}^\ell(M,\omega)$ with respect to the C^ℓ-topology. Moreover, $\mathcal{J}^\ell_{\mathrm{reg}} \subset \mathcal{J}^\ell_{\mathrm{reg},K}$ and hence, by the first part of the proof, $\mathcal{J}^\ell_{\mathrm{reg},K}$ is also dense in $\mathcal{J}^\ell(M,\omega)$. This implies that $\mathcal{J}_{\mathrm{reg},K}$ is dense in $\mathcal{J}$ with respect to the C^ℓ topology. To see this let $J \in \mathcal{J}$, approximate it in the C^ℓ-topology by an element $J' \in \mathcal{J}^\ell_{\mathrm{reg},K}$, and then approximate J' in the C^ℓ-topology by an element $J'' \in \mathcal{J}^\ell_{\mathrm{reg},K} \cap \mathcal{J} = \mathcal{J}_{\mathrm{reg},K}$. Thus we have proved that the set $\mathcal{J}_{\mathrm{reg},K}$ is dense in $\mathcal{J}$ with respect to the C^ℓ-topology. But this holds for any ℓ and hence $\mathcal{J}_{\mathrm{reg},K}$ is dense in $\mathcal{J}$ with respect to the C^∞-topology. In fact, given $J \in \mathcal{J}$ choose a sequence $J_\nu \in \mathcal{J}_{\mathrm{reg},K}$ such that $\|J - J_\nu\|_{C^\nu} \leq 2^{-\nu}$. Then J_ν converges to J in the C^∞-topology. Thus $\mathcal{J}_{\mathrm{reg}}$ is the intersection of the countable number of open dense sets $\mathcal{J}_{\mathrm{reg},K}$, and so has second category as required. $\square$

This completes the proof of Theorem 3.1.2. Theorem 3.1.3 is proved in a similar way: we take a path in the connected space $\mathcal{J}$ which joins J_0 to J_1 and then perturb it to be transverse to π. This is possible by the Sard-Smale theorem quoted before. Arguments analogous to those above (applied now to the space of paths) show that the perturbed path can be chosen to be C^∞-smooth.

Remark 3.4.2 The above proof shows that a point $(u, J) \in \mathcal{M}(A, J)$ is regular in the sense of Definition 3.1.1 if and only if the operator

$$D_u : W^{k,p}(u^*TM) \to W^{k-1,p}(\Lambda^{0,1}T^*\Sigma \otimes_J u^*TM)$$

is surjective for some choice of k, p. Hence $d\pi(u, J)$ is surjective at this point. By the implicit function theorem this implies that any smooth path $[0,1] \to \mathcal{J} : t \mapsto J_t$ which starts at $J_0 = J$ can be lifted, on some interval $[0, \varepsilon)$, to a path $[0, \varepsilon) \to \mathcal{M}(A, \mathcal{J}) : t \mapsto (u_t, J_t)$ in the universal moduli space which starts at $u_0 = u$. $\square$

3.5   A regularity criterion

We close this chapter by establishing a simple criterion for an almost complex structure J to be regular. In [27] Grothendieck proved that any holomorphic bundle E over $S^2 = \mathbb{C}P^1$ is holomorphically equivalent to a sum of holomorphic line bundles. Moreover, this splitting is unique up to the order of the summands. Hence $E = L_1 \oplus \cdots \oplus L_n$ is completely characterized by the set of Chern classes $c_1(L_1), \ldots, c_1(L_n)$. Note that the sum $c_1(E) = \sum_i c_1(L_i)$ is a topological invariant, but that the set $c_1(L_1), \ldots, c_1(L_n)$ may vary as $u : S^2 \to M$ varies in a connected component of the space of J-holomorphic A-spheres. In what follows we will identify the Chern class $c_1(L)$ with the corresponding Chern number $\langle c_1(L), [S^2] \rangle$.

Lemma 3.5.1 *Assume J is integrable and let $u : \mathbb{C}P^1 \to M$ be a J-holomorphic curve. Suppose that every summand of u^*TM has Chern number $c_1 \geq -1$. Then D_u is onto.*

Proof: If J is integrable then the operator D_u is exactly the Dolbeault derivative $\bar{\partial}$, and so it is surjective if and only if the Dolbeault cohomology group $H^{0,1}_{\bar{\partial}}(u^*TM)$ vanishes. But for any holomorphic line bundle L

$$H^{0,1}_{\bar{\partial}}(\mathbb{C}P^1, L) \cong (H^{1,0}_{\bar{\partial}}(\mathbb{C}P^1, L^*))^*.$$

But $H^{1,0}_{\bar{\partial}}(\mathbb{C}P^1, L^*)$ is just the space of holomorphic 1-forms with values in the dual bundle L^* and so is isomorphic to the space $H^0(\mathbb{C}P^1, \mathcal{O}(L^* \otimes K))$ of holomorphic sections of the bundle $L^* \otimes K$ where $K = T^*\mathbb{C}P^1$ is the canonical bundle. This is an easy special case of Kodaira-Serre duality (cf. [25, Ch 1 §2]) which can be checked directly by considering the transition maps. But, by the Kodaira vanishing theorem, a line bundle L' on $\mathbb{C}P^1$ has no holomorphic sections if and only if $c_1(L') < 0$. Thus D_u is surjective if and only if $c_1(L^* \otimes K) < 0$. But $c_1(L^* \otimes K) = -c_1(L) - 2$ and so this is equivalent to $c_1(L) > -2$. $\square$

Lemma 3.5.2 *Let J be an integrable complex structure on a 4-dimensional manifold M and $u : \mathbb{C}P^1 \to M$ be an embedded J-holomorphic curve. Then D_u is onto if and only if $c_1(u^*TM) \geq 1$.*

Proof: If $u \in \mathcal{M}(A, J)$ is an embedded J-holomorphic curve with image $u(S^2) = C$ then the pullback tangent bundle u^*TM splits into complex subbundles

$$u^*TM \cong T_C M = TC \oplus \nu$$

where $\nu \to C$ is a normal bundle with respect to some Hermitian structure. The Chern number of TC is 2 and, since C is embedded, the Chern number of ν is $p = C \cdot C$. Hence

$$c_1(u^*TM) = 2 + C \cdot C = 2 + p.$$

Now, according to Grothendieck, the bundle $u^*TM \to S^2$ splits into holomorphic subbundles

$$u^*TM = L_1 \oplus L_2$$

whose Chern classes $k_i = c_1(L_i)$ satisfy

$$k_1 + k_2 = 2 + p.$$

Moreover, the tangent bundle TC of C is a holomorphic subbundle of $L_1 \oplus L_2$. Therefore, it is either one of the summands, or it maps in a non-trivial way to both of L_1 and L_2. In the first case the other summand has Chern number p and so both factors have Chern number $k_i \geq -1$. Hence it follows from Lemma 3.5.1 that the pair (u, J) is regular. In the second case there is a non-trivial holomorphic section of the line bundle $\mathrm{Hom}(TC, L_i) = T^*C \otimes L_i$ for $i = 1, 2$. This is possible if and only if $c_1(T^*C \otimes L_i) = k_i - 2 \geq 0$ for $i = 1, 2$ since line bundles with negative Chern number have no non-zero sections. Hence it follows again from Lemma 3.5.1 that the pair (u, J) is regular. Moreover, note that this second case can occur only when $p \geq 2$. This proves the lemma. $\square$

Remark 3.5.3 Let $C = u(\Sigma)$ be the image of a J-holomorphic curve $u : \Sigma \to M$ in a symplectic 4-manifold M. The adjunction formula of [44] asserts that the *virtual genus* given by

$$2g(C) - 2 = C \cdot C - \langle c_1, [C] \rangle$$

is greater than or equal to the genus of Σ

$$g(C) \geq g(\Sigma)$$

with equality if and only if C is embedded. In particular this implies that if a homology class $A \in H_2(M)$ can be represented by an embedded J-holomorphic curve $u : \Sigma \to M$ then every other J-holomorphic curve $v : \Sigma \to M$ which represents A must also be embedded. In other words, one cannot deform an embedded J-holomorphic curve in a 4-manifold in such a way as to produce a singularity. $\square$

Corollary 3.5.4 *Let J be an integrable complex structure on a 4-dimensional manifold M, and consider an embedded J-holomorphic sphere C with self-intersection number $C \cdot C = p$. Then J is regular for the class $A = [C]$ if and only if $p \geq -1$.*

Proof: Lemma 3.5.2 and Remark 3.5.3. $\square$

Observe that this is the first place in our discussion where it is important to restrict to spheres. Our other results apply to J-holomorphic curves with fixed domain (Σ, j), but these are almost never regular because, even when J is integrable, the condition $H^{0,1}_{\bar\partial}(\Sigma, u^*TM) = 0$ is usually not satisfied. The point is that when J varies on M one must usually vary j on Σ in order to find a corresponding deformation of the curve. Thus, one has to set up the Fredholm theory in such a way that j is allowed to vary in Teichmüller space. This presents no problem at this stage, but it does make the discussion of compactness more complicated, since Teichmüller space, even when quotiented out by the mapping class group, is not compact.

Example 3.5.5 Suppose that M is the product of S^2 with a Kähler manifold V. Thus both $\omega = \omega_1 \times \omega_2$ and $J = J_1 \times J_2$ respect the product structure. If A is the homology class represented by the spheres $S^2 \times \{\mathrm{pt}\}$ then the J-holomorphic A-spheres are given by the maps $z \mapsto (\phi(z), x_0)$ where ϕ is a fractional linear transformation and $x_0 \in V$. Hence the above lemma implies that $J \in \mathcal{J}_{\mathrm{reg}}(A)$. $\square$

So far, there are no general methods known for proving that a given non-integrable J is regular for the class A. However, in the case of 4-manifolds one can do this by using positivity of intersections provided that $c_1(A) > 0$: see Hofer–Lizan–Sikorav [31] and Lorek [40]. When extending these results to curves of higher genus one has to be somewhat careful because the question of whether a line bundle has holomorphic sections is no longer purely topological, depending on the Chern number alone. Other respects in which moduli spaces of curves of higher genus differ from those of spheres are discussed in [47].

Chapter 4

Compactness

Because any manifold V is cobordant to the empty manifold via the non-compact cobordism $V \times [0, 1)$, Theorem 3.1.3 is useless unless one can establish some kind of compactness. Now in the case $\Sigma = S^2$ the manifold $\mathcal{M}(A, J)$ itself cannot be compact (unless it is empty) since the non-compact group $G = \mathrm{PSL}(2, \mathbb{C})$ of biholomorphic maps of S^2 acts on this space by reparametrization. However, in some cases the space $\mathcal{M}(A, J)/G$ of unparametrized spheres is compact.

If J is tamed by a symplectic form ω then it follows from the energy identity (see Lemma 4.1.2 below) that there is a uniform bound on the $W^{1,2}$-norm of all J-holomorphic curves in a given homology class. This is the Sobolev borderline case $kp = 2$ and, as a result, the space of such curves will in general not be compact. This is due to the conformal invariance of the energy in 2 dimensions and leads to the phenomenon of bubbling, which was first discovered by Sacks and Uhlenbeck in the context of minimal surfaces [71]. In fact, it follows from the usual elliptic bootstrapping argument that any sequence u_ν in $\mathcal{M}(A, J)$ which is bounded in the $W^{1,p}$-norm for some $p > 2$ has a subsequence which converges uniformly with all derivatives. On the other hand if the first derivatives of u_ν are only bounded in L^2 but not in L^p a simple geometric argument using conformal rescaling allows one to construct a J-holomorphic map $v : \mathbb{C} \to M$ with finite area which, by "removal of singularities", can be extended to $S^2 = \mathbb{C} \cup \{\infty\}$. This is the phenomenon of "bubbling off of spheres". Sometimes, by choosing the class A carefully, one can show that this cannot happen. In this case, the space $\mathcal{M}(A, J)/G$ is compact. In other cases, bubbling off can occur, and one must proceed with more care.

The method of proof used below is very close to that followed by Floer. Gromov's original approach was somewhat different. He argued geometrically, using isoperimetric inequalities and the Schwartz Lemma for conformal maps. More details of his proofs have been written up by Pansu [61]. The flavour of his arguments may be sampled in the proof of Theorem 4.2.1 (removal of singularities) below. Other accounts of this subject can be found in [32], [62], [73], and [88]. In Section 4.3 we shall prove the simplest version of the compactness theorem. This gives a criterion for the moduli space of unparametrized curves to be compact. The criterion is that A is minimal in the sense that there is no other class B with $0 < \omega(B) < \omega(A)$. In particular this implies that A is not a multiple class so that the space $\mathcal{M}(A, J)$ contains all J-holomorphic A-curves. If this is not satisfied, one can form a compactification by adding suitable "cusp-curves" and this will be discussed in detail in Sections 4.4 and 4.5. The terminology cusp-curve was introduced by Gromov. However, these curves would be called reducible curves in the setting of algebraic geometry.

We will assume, for simplicity, that the Riemann surface Σ is a sphere. In fact, the arguments go through without essential change to the general case provided that one fixes the complex structure j on Σ. Note that in all cases, the bubbles which appear are spherical. We will denote the reparametrization group by G. Thus G is the non-compact group $\text{PSL}(2,\mathbb{C})$ acting on $S^2 = \mathbb{C} \cup \{\infty\}$ by fractional linear transformations

$$\phi_A(z) = \frac{az+b}{cz+d}, \qquad A = \begin{pmatrix} a & b \\ c & d \end{pmatrix} \in \text{SL}(2,\mathbb{C}). \qquad (4.1)$$

We begin with a discussion of metrics on M which are related to J and ω. We will assume that $J : TM \to TM$ is an ω-**tame** almost complex structure on M. This means that

$$\omega(v, Jv) > 0 \qquad \text{for} \qquad v \neq 0.$$

This condition is weaker than compatibility since we no longer require the bilinear form $(v, w) \mapsto \omega(v, Jw)$ to be symmetric. However there is an induced Riemannian metric

$$g_J(v, w) = \langle v, w \rangle_J = \tfrac{1}{2} \left(\omega(v, Jw) + \omega(w, Jv) \right).$$

Henceforth, we assume that M is provided with this metric. It is easy to check that the endomorphism J is an isometry and is skew self-adjoint with respect to this metric. It follows that a J-holomorphic map u is conformal with respect to the Poincaré metric on (Σ, j) and the above metric on M.

4.1 Energy

We define the **energy** of a smooth map $u : \Sigma \to M$ to be the L^2-norm of the 1-form $du \in \Omega^1(u^*TM)$:

$$E(u) = \frac{1}{2} \int_\Sigma |du|_J^2 \, dA.$$

Here the norm $|du|_J$ is not well defined unless we fix a metric on Σ. However, the following exercise shows that the product $|du|_J^2 \, dA$ is independent of the choice of this metric.

Exercise 4.1.1 Let Σ be a Riemann surface with complex structure j, $E \to \Sigma$ be a Riemannian vector bundle, and $\zeta \in \Omega^1(E)$ be a 1-form. Choose a conformal coordinate chart $\alpha : U_\alpha \to \mathbb{C}$ on Σ and an orthogonal trivialization and denote by $z = s + it = \alpha(p)$ the coordinates of a generic point $p \in U_\alpha \subset \Sigma$. Moreover, choose a local orthogonal trivialization $\phi_\alpha : \pi^{-1}(U_\alpha) \to \alpha(U_\alpha) \times \mathbb{R}^n$ of E such that $p_1 \circ \phi_\alpha = \alpha \circ \pi$. In these coordinates the 1-form ζ can be represented by the form

$$\zeta^\alpha = \xi^\alpha ds + \eta^\alpha dt \in \Omega^1(\alpha(U_\alpha), \mathbb{R}^n)$$

where $p_2 \circ \phi_\alpha \circ \zeta = \alpha^* \zeta^\alpha$. Prove that the local expressions

$$|\zeta|^2 dA := \alpha^* \left(\left(|\xi^\alpha|^2 + |\eta^\alpha|^2 \right) \, ds \wedge dt \right)$$

on U_α determine a well defined global 2-form on Σ. Its integral is the L^2-norm of ζ. $\qquad\Box$

Lemma 4.1.2 *If J is ω-tame then the energy identity*

$$E(u) = \int_\Sigma u^* \omega \qquad (4.2)$$

holds for all J-holomorphic curves u.

Proof: Exercise. $\qquad\Box$

Next we give a precise statement of the basic compactness result which is proved by "elliptic bootstrapping". In this theorem the domain Σ may be noncompact. The proof will be carried out in Appendix B.

Theorem 4.1.3 *Let $u_\nu : \Sigma \to M$ be a sequence of J-holomorphic curves such that*

$$\sup_\nu \|du_\nu\|_{L^\infty(K)} < \infty$$

for every compact subset $K \subset \Sigma$. Then u_ν has a subsequence which converges uniformly with all derivatives on compact subsets of Σ.

The conclusion of the previous theorem continuous to hold when the maps u_ν are uniformly bounded in the $W^{1,p}$-norm for some $p > 2$. Now, if the elements of the sequence u_ν all represent the same homology class then the energy identity of the Lemma 4.1.2 guarantees a uniform bound for the $W^{1,2}$-norm. However, the compactness theorem breaks down in the case $p = 2$ and the conformal invariance of the energy leads to the phenomenon of *bubbling*. An important ingredient in understanding this phenomenon is the removable singularity theorem for J-holomorphic curves of finite energy which we will discuss next.

4.2 Removal of Singularities

The removable singularity theorem says that any J-holomorphic curve $u : B - \{0\} \to M$ on the punctured disc which has finite energy extends smoothly to the disc.

Theorem 4.2.1 (Removal of singularities) *Let J be a smooth ω-tame almost complex structure on a compact manifold M with associated metric g_J. If $u : B - \{0\} \to M$ is a J-holomorphic curve with finite energy $E(u) < \infty$ then u extends to a smooth map $B \to M$.*

Proof of continuity: Here we follow essentially the line of argument in Gromov's original work. We assume for simplicity that J is compatible with ω so that the J-holomorphic curves minimize the energy. As pointed out by Pansu in [61, §37], one can prove the removable singularity theorem in this case by using the monotonicity theorem for minimal surfaces. This states that there are constants $c > 0$ and $\varepsilon_0 > 0$ (which depend on M and the metric g_J) such that for every minimal surface S in (M, g_J) which goes through the point x

$$\text{area}_{g_J}(S \cap B(x, \epsilon)) \geq c\epsilon^2$$

for $0 < \varepsilon < \epsilon_0$. (See [37, 3.15].) To apply this, suppose that $u(z)$ has two limit points p and q as $z \to 0$. If δ is chosen to be less than $d(p, q)/3$, then the monotonicity theorem implies that each connected component of $u^{-1}(B(p, \delta))$ which meets $u^{-1}(B(p, \delta/2))$ is taken by u to a surface in M which has area $\geq c\delta^2/4$. Therefore, because $\text{Im}\, u$ is minimal and has finite area (or energy) $E(u)$ by (4.2), there can only be a finite number of such components. Similar remarks apply to q. Hence there exists an $r_0 > 0$ such that, for any $r < r_0$, the image γ_r of the circle $\{z \in \mathbb{C} \mid |z| = r\}$ under u meets both $B(p, \delta/2)$ and $B(q, \delta/2)$, and so must have length $\ell(\gamma_r) > \delta$. But then, the conformality of u implies that $|du| = \frac{1}{r}|\partial u/\partial\theta|$, and we find that

$$\begin{aligned}
E(u) &= \int_{(0,1]\times S^1} \frac{|\partial u/\partial\theta|^2}{r^2} r\, dr \wedge d\theta \\
&\geq \int_{(0,1]} \left[\int_{S^1} |\partial u/\partial\theta|\, d\theta\right]^2 \frac{1}{2\pi r}\, dr \\
&= \int_{(0,1]} \frac{\ell(\gamma_r)^2}{2\pi r}\, dr \\
&\geq \int_{(0,r_0]} \frac{\delta^2}{2\pi r}\, dr,
\end{aligned}$$

which is impossible because $E(u)$ is finite. The geometric idea here is that, because of the conformality, if the loops γ_r are long they must also stretch out in the radial direction, and hence form a surface of infinite area. $\qquad\square$

We now give a proof of the smoothness of the extension which relies on the following **a priori estimate**.

Lemma 4.2.2 (A priori estimate) *Assume M is compact and J is a smooth ω-tame almost complex structure. Then there exists a constant $\hbar > 0$ such that the following holds. If $r > 0$ and $u : B_r \to M$ is a J-holomorphic curve such that*

$$E(u; B_r) = \int_{B_r} |du|^2 < \hbar$$

then

$$|du(0)|^2 \leq \frac{8}{\pi r^2} \int_{B_r} |du|^2.$$

The proof relies on a partial differential inequality of the form $\Delta e \geq -Ae^2$ for the *energy density* $e = |du|^2$. The details are carried out in [73] for example. As a consequence of this estimate we obtain the following isoperimetric inequality which plays a crucial role in the proof of Theorem 4.2.1 and is related to the monotonicity property of minimal surfaces used above.

Lemma 4.2.3 (Isoperimetric inequality) *Assume that M is compact and that J is a smooth ω-tame almost complex structure. Let $u : B - \{0\} \to M$ be a J-holomorphic curve on the punctured disc with finite energy. Then there exist constants $\delta > 0$ and $c > 0$ such that*

$$E(u; B_r) \le c\,\ell(\gamma_r)^2, \qquad 0 < r < \delta$$

where $\gamma_r(\theta) = u(re^{i\theta})$ and $\ell(\gamma_r)$ denotes the length of the loop γ_r.

Proof: In view of Lemma 4.2.2 we have

$$|du(re^{i\theta})|^2 \le \frac{8}{\pi r^2}\varepsilon(2r), \qquad \varepsilon(r) = E_{B_r}(u)$$

for $r \le \delta$. Now the derivative of γ_r has norm $|\dot{\gamma}_r(\theta)| = r\,|du(re^{i\theta})|$. Hence the length of the loop γ_r is given by

$$\ell(\gamma_r) = r \int_0^{2\pi} |du(re^{i\theta})|\, d\theta \le \sqrt{32\pi\varepsilon(2r)}.$$

Hence $\ell(\gamma_r)$ converges to zero with r.

Now every sufficiently short loop $\gamma : S^1 \to M$ has a unique *local extension* $u_\gamma : B \to M$ defined by $u_\gamma(re^{i\theta}) = \exp_{\gamma(0)}(r\xi(\theta))$ where $\xi(\theta) \in T_{\gamma(0)}M$ is determined by the condition $\exp_{\gamma(0)}(\xi(\theta)) = \gamma(\theta)$. The area of this disc is the **local symplectic action** of γ and is bounded by the square of the length of γ. In other words, there exists a constant $c > 0$ such that for every sufficiently short loop $\gamma : S^1 \to M$

$$a(\gamma) = \int u_\gamma^* \omega \le c\,\ell(\gamma)^2$$

Denote by $u_r = u_{\gamma_r} : B_1 \to M$ the extension of the loop γ_r and consider the sphere $v_{\rho r} : S^2 \to M$ obtained from $u|_{B_r - B_\rho}$ with the boundary circles γ_ρ and γ_r filled in by the discs u_ρ and u_r. This sphere is contractible. It is the restriction of the smooth map $B_1 \times [\rho, r] \to M : (z, \tau) \mapsto u_\tau(z)$ to the boundary of the cylinder. Hence

$$E(u; B_r - B_\rho) + \int_{B_1} u_\rho^* \omega = \int_{B_1} u_r^* \omega$$

for $0 < \rho < r$ for r sufficiently small. Take the limit $\rho \to 0$ to obtain the required inequality $E(u; B_r) = \int u_r^* \omega \le c\,\ell(\gamma_r)^2$. $\qquad\square$

Proof of Theorem 4.2.1: Continue the notation of Lemma 4.2.3. In particular

$$\varepsilon(r) = E(u; B_r) = \int_{B_r} |du|^2 = \int_0^r \rho \int_0^{2\pi} |du(\rho e^{i\theta})|^2 \, d\rho d\theta.$$

It follows from the isoperimetric inequality of Lemma 4.2.3 that for r sufficiently small

$$
\begin{aligned}
\varepsilon(r) &\leq c\,\ell(\gamma_r)^2 \\
&= cr^2 \left(\int_0^{2\pi} |du(re^{i\theta})|\, d\theta \right)^2 \\
&\leq 2\pi cr^2 \int_0^{2\pi} |du(re^{i\theta})|^2\, d\theta \\
&= 2\pi cr\dot{\varepsilon}(r).
\end{aligned}
$$

With $\mu = 1/2\pi c$ this can be rewritten as $\mu/r \leq \dot{\varepsilon}(r)/\varepsilon(r)$. Integrating this inequality from r to r_1 we obtain $(r_1/r)^\mu \leq \varepsilon(r_1)/\varepsilon(r)$ and hence

$$
\varepsilon(r) \leq c_1 r^\mu
$$

where $c_1 = r_1^{-\mu}\varepsilon(r_1)$. For ρ sufficiently small the estimate of Lemma 4.2.2 shows that

$$
|du(\rho e^{i\theta})|^2 \leq \frac{8}{\pi\rho^2}\varepsilon(2\rho) \leq \frac{c_2}{\rho^{2-\mu}} \tag{4.3}
$$

with a suitable constant $c_2 > 0$ which is independent of ρ and θ. With $2 < p < 4/(2 - \mu)$ this implies

$$
\begin{aligned}
\int_{B_r} |du|^p &= \int_0^r \int_0^{2\pi} \rho|du(\rho e^{i\theta})|^p\, d\theta d\rho \\
&\leq c_3 \int_0^r \rho^{1-p(1-\mu/2)}\, d\rho.
\end{aligned}
$$

Since $p < 4/(2 - \mu)$ we have $1 - p(1 - \mu/2) > -1$ and hence the integral is finite. Hence $u \in W^{1,p}$ with $p > 2$. Moreover, it follows from (4.3) that u is Hölder continuous. Since M is compact this implies (again) that u extends to a continuous map on the closed unit disc. Now it follows from elliptic regularity in a local coordinate chart that u is smooth on the closed unit disc (see Proposition 3.2.2). □

This completes the proof of the removable singularity theorem. A similar result is true for boundary points of J-holomorphic curves: see Oh [59].

Exercise 4.2.4 Prove directly from (4.3) that u is Hölder continuous with exponent $\mu/2$. This is needed to prove that $u(B_r - \{0\})$ lies in a single coordinate chart for $r > 0$ sufficiently small. □

4.3　Bubbling

The next theorem shows how bubbles appear. We shall begin by discussing some notation. The most convenient way of describing holomorphic spheres is to identify S^2 with $\mathbb{C}P^1 = \mathbb{C} \cup \{\infty\}$. If S^2 denotes the unit sphere in $\mathbb{R}^3$ then such an identification is given by stereographic projection. Different choices of a stereographic projection correspond to the action of $SO(3) \simeq SU(2)/\{\pm 1\} \subset PSL(2, \mathbb{C})$ on $\mathbb{C}P^1 = \mathbb{C} \cup \{\infty\}$. With this identification a J-holomorphic sphere $u \in \mathcal{M}(A, J)$ is a smooth J-holomorphic curve $u : \mathbb{C} \to M$ such that the map $\mathbb{C} - \{0\} \to M : z \mapsto u(1/z)$ extends to a smooth map on $\mathbb{C}$. This class of maps is preserved under composition with fractional linear transformations $\phi_A : \mathbb{C} \cup \{\infty\} \to \mathbb{C} \cup \{\infty\}$. A sequence of J-holomorphic curves $u_\nu : \mathbb{C} \to M$ is said to converge on $\mathbb{C} \cup \{\infty\}$ if both $u_\nu(z)$ and $u_\nu(1/z)$ converge uniformly with all derivatives on compact subsets of $\mathbb{C}$. We make use of the following estimate on the derivatives.

Lemma 4.3.1 *If* $u : \mathbb{C} \to M$ *represents a smooth map* $S^2 \to M$ *then there exists a constant* $c > 0$ *such that*

$$|du(z)| \leq \frac{c}{1 + |z|^2}$$

for $z \in \mathbb{C}$.

Proof: Define $v(z) = u(1/z)$ and note that the map $\phi(z) = 1/z$ satisfies $|d\phi(z)| = |z|^{-2}$. Hence $|du(z)| = |z|^{-2}|dv(1/z)|$ and this proves the estimate for $|z| \geq 1$ with $c = \|dv\|_{L^\infty(B_1)}$. For $|z| \leq 1$ the estimate obviously holds with $c = \|du\|_{L^\infty(B_1)}$. $\square$

Theorem 4.3.2 *Assume that there is no spherical homology class[1]* $B \in H_2(M)$ *such that* $0 < \omega(B) < \omega(A)$. *Then the moduli space* $\mathcal{M}(A, J)/G$ *is compact.*

Proof: Let $u_\nu : \mathbb{C} \to M$ be a sequence of J-holomorphic curves which represent the class A. We must prove that there exists a sequence $A_\nu \in PSL(2, \mathbb{C})$ such that $u_\nu \circ \phi_{A_\nu}$ has a subsequence which converges on $\mathbb{C} \cup \{\infty\}$.

It follows from Lemma 4.3.1 that $|du_\nu(z)|$ assumes its maximum at some point $a_\nu \in \mathbb{C}$. Denote

$$c_\nu = |du_\nu(a_\nu)| = \|du_\nu\|_{L^\infty}$$

and define the reparametrized curve $v_\nu : \mathbb{C} \to M$ by

$$v_\nu(z) = u_\nu(a_\nu + c_\nu^{-1}z).$$

This curve satisfies

$$|dv_\nu(0)| = 1, \qquad \|dv_\nu\|_{L^\infty} \leq 1, \qquad E(v_\nu) = E(u_\nu) = \omega(A).$$

The last identity follows from the conformal invariance of the energy and the formula (4.2). By Theorem 4.1.3 there exists a subsequence, still denoted by v_ν, which converges uniformly with all derivatives on compact sets. The limit function $v : \mathbb{C} \to M$ is again a J-holomorphic curve such that

$$|dv(0)| = 1, \qquad 0 < E(v) = \int_{\mathbb{C}} v^*\omega \leq \omega(A).$$

Now the removable singularity theorem implies that the map $\mathbb{C} - \{0\} \to M : z \mapsto v(1/z)$ extends to a smooth map on $\mathbb{C}$ and hence v represents a J-holomorphic sphere.

---

[1] A homology class $B \in H_2(M)$ is called **spherical** if it is in the image of the Hurewicz homomorphism $\pi_2(M) \to H_2(M)$.

In order to show that $\mathcal{M}(A, J)/G$ is compact, we must prove that the functions $v'_\nu(z) = v_\nu(1/z)$ converge to $v(1/z)$ uniformly on compact neighbourhoods of 0. Assume otherwise that

$$c'_\nu = \|dv'_\nu\|_{L^\infty} \to \infty,$$

passing to a subsequence if necessary. As before, let $a'_\nu \in \mathbb{C}$ be the point at which $|dv'_\nu(z)|$ attains its maximum. Since $v'_\nu(z)$ converges to $v(1/z)$ uniformly on compact subsets of $\mathbb{C} - \{0\}$ it follows that $a'_\nu \to 0$. Consider the reparametrized curves

$$w_\nu(z) = v'_\nu(a'_\nu + c'_\nu{}^{-1}z).$$

These maps again satisfy

$$|dw_\nu(0)| = 1, \qquad \|dw_\nu\|_{L^\infty} \leq 1, \qquad E(w_\nu) = \omega(A).$$

Hence, passing to a further subsequence, we may assume that w_ν converges uniformly with all derivatives on compact sets to a J-holomorphic curve $w : \mathbb{C} \to M$ such that

$$|dw(0)| = 1, \qquad 0 < E(w) = \int_{\mathbb{C}} w^*\omega \leq \omega(A).$$

Again it follows from the removable singularity theorem that $w(1/z)$ extends to a smooth J-holomorphic curve at 0 and hence represents a J-holomorphic sphere.

Now let B and C denote the homology classes represented by v and w, respectively. Then $\omega(B)$ and $\omega(C)$ are positive and we derive the desired contradiction by showing that

$$\omega(B) + \omega(C) \leq \omega(A).$$

This holds because v and w are limits of "disjoint pieces" of the sequence u_ν. More precisely, denote by $B_R = B_R(0) \subset \mathbb{C}$ denotes the ball of radius R centered at 0 and by

$$E(w; \Omega) = \int_\Omega w^*\omega$$

the energy of the J-holomorphic curve w on the domain $\Omega \subset \mathbb{C}$. Then for every $\varepsilon > 0$ we have

$$
\begin{aligned}
\omega(C) &= \lim_{R \to \infty} E(w; B_R) \\
&= \lim_{R \to \infty} \lim_{\nu \to \infty} E(w_\nu; B_R) \\
&= \lim_{R \to \infty} \lim_{\nu \to \infty} E(v'_\nu; B_{Rc'_\nu{}^{-1}}(a'_\nu)) \\
&\leq \lim_{R \to \infty} \lim_{\nu \to \infty} E(v'_\nu; B_\varepsilon) \\
&= \lim_{\nu \to \infty} (E(v_\nu) - E(v_\nu; B_{1/\varepsilon})) \\
&= \omega(A) - E(v; B_{1/\varepsilon}).
\end{aligned}
$$

Take the limit $\varepsilon \to 0$ to obtain the required inequality $\omega(C) \leq \omega(A) - \omega(B)$. Note that $\omega(B) + \omega(C)$ need not equal $\omega(A)$ because there may be yet other bubbles which we have not detected. Different choices of the center and size of the rescaling may give rise to different limits. $\qquad\square$

A simple modification of the above proof with varying almost complex structures gives the following result.

Corollary 4.3.3 *Assume that there is no spherical homology class $B \in H_2(M)$ such that $0 < \omega(B) < \omega(A)$. Assume that J_ν is a sequence of almost complex structures which converges to J in the C^∞-topology. Let $u_\nu : \mathbb{C} \cup \{\infty\} \to M$ be a sequence of J_ν-holomorphic A-spheres. Then there exist matrices $A_\nu \in \mathrm{SL}(2, \mathbb{C})$ such that $u_\nu \circ \phi_{A_\nu}$ has a convergent subsequence.*

We say that the class A is **indecomposable** if it does not decompose as a sum $A = A_1 + \ldots + A_k$ of classes which are spherical and satisfy $\omega(A_i) > 0$ for all i. Note that A may be an indecomposable class without satisfying the assumption of Theorem 4.3.2. For indecomposable classes the conclusion of Theorem 4.3.2 remains valid but the proof requires the more sophisticated compactness theorem discussed in the next section. Here we only state the result.

Theorem 4.3.4 *If A is indecomposable then the moduli space $\mathcal{M}(A, J)/\mathrm{G}$ of unparametrized J-holomorphic A-spheres is compact for all ω-compatible J.*

Example 4.3.5 Let $M = S^2 \times S^2$ with symplectic form $\omega_\lambda = \lambda\omega_1 \times \omega_2$ where each ω_i is an area form on the sphere with total area π and where $\lambda \geq 1$. Let $A = [S^2 \times pt]$ and $B = [pt \times S^2]$. If $\lambda = 1$, then $\omega_\lambda(A) = \pi$ is the smallest positive value taken by $[\omega_\lambda]$ on $\pi_2(M)$. Thus A is indecomposable and $\mathcal{M}(A, J)/\mathrm{G}$ is compact. But if $\lambda > 1$, ω_λ is positive on the class $A - B$ of the anti-diagonal $\{(z, \alpha(z)) : z \in S^2\}$ (where α is the antipodal map), and it becomes possible for the set of J-holomorphic A-spheres to be non-compact. If J is a product, $\mathcal{M}(A, J)/\mathrm{G}$ is compact, but there are ω_λ-compatible J for which it is not. For example, one can take J to be Hirzebruch's complex structure on $S^2 \times S^2$ which is obtained by identifying $S^2 \times S^2$ with the projectivization of the rank 2 complex vector bundle over $S^2 = \mathbb{C}P^1$ which has first Chern class equal to 2. However, one can show that, no matter what λ is, the moduli space $\mathcal{M}(A, J)/\mathrm{G}$ is compact for a generic J. $\square$

Exercise 4.3.6 Let $M = S^2 \times \mathbb{T}^{2n-2}$ be the product of the 2-sphere with a torus with symplectic form $\omega = \omega_1 \times \omega_2$, and let $A = [S^2 \times \{pt\}]$. Show that A is indecomposable. $\square$

When one applies these results to get information about the symplectic manifold, the most useful tool is the **evaluation map**. If the curves under consideration are spheres, then the domain of this map is the quotient space $\mathcal{M}(A, J) \times_\mathrm{G} S^2$ where the reparametrization group G acts on the product diagonally

$$\phi \cdot (u, z) = (u \circ \phi^{-1}, \phi(z)).$$

The evaluation map $e = e_A = e_{A,J} : \mathcal{M}(A, J) \times_\mathrm{G} S^2 \to M$ is given by the formula

$$e(u, z) = u(z).$$

Theorem 4.3.7 *Let A be a indecomposable class, and J_1, J_2 two elements of $\mathcal{J}_{\mathrm{reg}}$. Then the evaluation maps e_{A,J_1} and e_{A,J_2} are compactly bordant.*

Proof: The evaluation maps extend over $\mathcal{M}(A, \{J_\lambda\}_\lambda) \times_\mathrm{G} S^2$. $\square$

4.4 Gromov compactness

Throughout we assume that (M, ω) is a compact symplectic manifold and denote by $\mathcal{J} = \mathcal{J}_\tau(M, \omega)$ the space of smooth ω-tame almost complex structures on M. If $J \in \mathcal{J}$ and $A \in H_2(M, \mathbb{Z})$ is not an indecomposable class then the unparametrized moduli space $\mathcal{C}(A, J) = \mathcal{M}(A, J)/G$ will in general not be compact. A sequence of A-curves may degenerate into a **cusp-curve** C. This is a connected union

$$C = C^1 \cup C^2 \cup \ldots \cup C^N$$

of J-holomorphic spheres C^j which are called components. Each component is parametrized by a smooth nonconstant J-holomorphic map $u^j : \mathbb{C}P^1 \to M$ which is not required to be simple. A special case occurs when $N = 1$ and the curve $u = u^1$ is multiply covered. Then the convergence may be uniform, with all derivatives, but the limit-curve u is not included in the space $\mathcal{M}(A, J)$ which only consists of simple curves. In [64] Ruan uses the term *reducible curve* for a collection $u = (u^1, \ldots, u^N)$ of nonconstant J-holomorphic curves with a connected image such that either $N \geq 2$ or $N = 1$ and $u = u^1$ is multiply covered. We shall adopt Gromov's terminology in [26] and use the term *cusp-curve*.

Remark 4.4.1 A cusp-curve $u = (u^1, \ldots, u^N)$ can be ordered such that the set $C^1 \cup \ldots \cup C^j$ with $C^k = u^k(\mathbb{C}P^1)$ is connected for every j. This means that there exist numbers

$$j_2, \ldots, j_N, \qquad 1 \leq j_k < k,$$

and points $w_k, z_k \in \mathbb{C}P^1$ such that

$$u^{j_k}(w_k) = u^k(z_k).$$

As a matter of normalization we may assume for example that $z_k = \infty$ for every k. $\square$

Remark 4.4.2 A cusp-curve $u = (u^1, \ldots, u^N)$ can be parametrized by a single smooth but not J-holomorphic map $v : \mathbb{C}P^1 \to M$. To see this order the curves $u^1, \ldots, u^N$ as in the previous remark and argue by induction as follows. Assume that $v^j : \mathbb{C}P^1 \to M$ has been constructed for $j \geq 1$ as to parametrize $C^1 \cup \ldots \cup C^j$. Assume without loss of generality that $v^j(\infty) = u^{j+1}(0)$ and choose a smooth map $v^{j+1} : \mathbb{C} \cup \{\infty\} \to M$ which covers $v^j(\mathbb{C}P^1)$ on the unit disc, maps the unit circle to $v^j(\infty) = u^{j+1}(0)$ and covers C^{j+1} on the complement of the unit disc. More explicitly, choose a smooth cutoff function $\beta : \mathbb{R} \to [0, 1]$ such that $\beta(r) = 1$ for $r \leq 1/2$ and $r \geq 2$ and $\beta(r) = 0$ for r close to 1, and define

$$v^{j+1}(z) = \begin{cases} v^j(\beta(|z|^2)^{-1}z), & \text{if } |z| < 1, \\ u^{j+1}(\beta(|z|^2)z), & \text{if } |z| > 1. \end{cases}$$

This curve parametrizes $C^1 \cup \ldots \cup C^{j+1}$. $\square$

A sequence of J-holomorphic curves $u_\nu : \mathbb{C}P^1 = \mathbb{C} \cup \{\infty\} \to M$ is said to **converge weakly** to a curve $u = (u^1, \ldots, u^N)$ (which may consist only of one component) if the following holds.

(i) For every $j \leq N$ there exists a sequence $\phi_\nu^j : \mathbb{C}P^1 \to \mathbb{C}P^1$ of fractional linear transformations and a finite set $X^j \subset \mathbb{C}P^1$ such that $u_\nu \circ \phi_\nu^j$ converges to u^j uniformly with all derivatives on compact subsets of $\mathbb{C}P^1 - X^j$.

(ii) There exists a sequence of orientation preserving (but not holomorphic) diffeomorphisms $f_\nu : \mathbb{C}P^1 \to \mathbb{C}P^1$ such that $u_\nu \circ f_\nu$ converges in the C^0-topology to a parametrization $v : \mathbb{C}P^1 \to M$ of the cusp-curve u as in Remark 4.4.2.

It follows from this definition that every point on the union $C = \bigcup_j u^j(\mathbb{C}P^1)$ is a limit of some sequence $p_\nu \in u_\nu(\mathbb{C}P^1)$. It follows also from the definition of weak convergence that the map $u_\nu : \mathbb{C}P^1 \to M$ is homotopic to the connected sum

$$u^1 \# u^2 \# \cdots \# u^N : \mathbb{C}P^1 \to M$$

for ν sufficiently large. Hence, in particular,

$$\omega(A_\nu) = \sum_{j=1}^{N} \omega(A^j), \qquad c_1(A_\nu) = \sum_{j=1}^{N} c_1(A^j),$$

where $A_\nu \in H_2(M, \mathbb{Z})$ is the homology class of u_ν and A^j is the homology class of u^j. In our applications below we shall in general assume that all the u_ν represent the same homology class $A \in H_2(M, \mathbb{Z})$ and it then follows from weak convergence that the connected sum $u^1 \# \cdots \# u^N$ also represents the class A. We point out that the notion of weak convergence also makes sense when the u_ν are not J-holomorphic.

Theorem 4.4.3 (Gromov's compactness) *Assume M is compact and let $J_\nu \in \mathcal{J}_\tau(M, \omega)$ be a sequence of ω-tame almost complex structures which converges to J in the C^∞-topology. Then any sequence $u_\nu : \mathbb{C}P^1 \to M$ of J_ν-holomorphic spheres with $\sup_\nu E(u_\nu) < \infty$ has a subsequence which converges weakly to a (possibly reducible) J-holomorphic curve $u = (u^1, \ldots, u^N)$.*

Various versions of this theorem are proved in [32], [61], [62], and [88]. The proof given in the next section essentially follows the line of argument in [32].

Corollary 4.4.4 *Given $K > 0$, every element $J \in \mathcal{J}$ has an open neighbourhood $\mathcal{N}(J)$ such that there are only finitely many classes A with $\omega(A) \leq K$ which have J'-holomorphic representatives for some $J' \in \mathcal{N}(J)$.*

Proof: If not, there would be a sequence of curves, each in a different homology class, of bounded energy. But these homology classes lie in the discrete lattice $H_2(M; \mathbb{Z})$, and so can have no convergent subsequence. $\square$

If u_ν is a weakly converging sequence, we will see in Lemma 4.5.5 that there is a finite set X such that u_ν converges to some J-holomorphic map u_∞ on compact subsets of $\mathbb{C}P^1 - X$. The corresponding component of the limiting cusp-curve can be thought of as basic one, the other components being "bubbles" at the points of X. When Σ is the sphere, this remark has little force, but if one considers a limit of maps u_ν whose domain is a *fixed* Riemann surface Σ, the basic component is the only one which can have genus > 0. Because there is no obvious holomorphic parametrization of the different components of a limiting cusp-curve, it is hard to define a good compactification for spaces $\mathcal{M}$ of parametrized curves. However, all that is needed to define symplectic invariants is to compactify the space $\mathcal{C}(A, J) = \mathcal{M}(A, J)/G$ of unparametrized curves. We shall describe such a compactification in the next chapter. This compactification can be understood without going through the detailed analysis involved in the proof of Gromov's compactness theorem and the reader may wish to go directly to Chapter 5.

4.5   Proof of Gromov compactness

One key step in the proof is the following result about the energy of a J-holomorphic curve on an arbitrarily long cylinder. It asserts that if the energy is sufficiently small then it cannot be spread out uniformly but must be concentrated near the ends. We phrase the result in terms of closed annuli $A(r, R) = B_R - \text{int } B_r$ for $r < R$. The relative size of the radii r, R will be all-important in what follows. One should think of the ratio R/r as being very large, and much bigger than e^T.

Lemma 4.5.1 *Let (M, ω) be a compact symplectic manifold and J be an ω-tame almost complex structure. Then there exist constants $c > 0$, $\hbar > 0$, and $T_0 > 0$ such that the following holds. If $u : A(r, R) \to M$ is a J-holomorphic curve such that $E(u) < \hbar$ then*

$$E(u; A(e^T r, e^{-T} R)) \le \frac{c}{T} E(u; r, R)$$

and

$$\int_0^{2\pi} \text{dist}\left(u(re^{T+i\theta}), u(Re^{-T+i\theta})\right) \, d\theta \le c\sqrt{\frac{E(u; A(r, R))}{T}}$$

for $T \ge T_0$.

Proof: Choose the constant $\hbar > 0$ as in Lemma 4.2.2 and consider the J-holomorphic curve $v(\tau + i\theta) = u(e^{\tau + i\theta})$ for $\log r < \tau < \log R$ and $\theta \in S^1 = \mathbb{R}/2\pi\mathbb{Z}$. Then for $\log r + T < \tau < \log R - T$, the projection $B_T \to [\log r, \log R] \times S^1$ is at most a T-fold cover and so we have $E(v; B_T(\tau + i\theta)) \le TE(v) = TE(u)$. Hence, by Lemma 4.2.2,

$$|dv(\tau + i\theta)|^2 \le \frac{8E(u)}{\pi T}, \qquad \log r + T < \tau < \log R - T.$$

If T is sufficiently large then the loop $\gamma_\tau(\theta) = v(\tau + i\theta)$ is sufficiently short and therefore has a local symplectic action $a(\gamma_\tau)$ as in the proof of Lemma 4.2.3. Recall that this action is defined as the integral of ω over a local disc $u_\tau : B_1 \to M$ with $u_\tau(e^{i\theta}) = \gamma_\tau(\theta)$ and satisfies

$$|a(\gamma_\tau)| \le c_1 \ell(\gamma_\tau)^2 \le c_2 \frac{E(u)}{T}$$

for $\log r + T < \tau < \log R - T$. Since u is a J-holomorphic curve we have

$$\frac{d}{d\tau} a(\gamma_\tau) = \int_0^{2\pi} |\dot{\gamma}_\tau(\theta)|^2 \, d\theta \ge \frac{\|\dot{\gamma}_\tau\|_{L^2}}{\sqrt{2\pi}} \ell(\gamma_\tau) \ge \frac{\|\partial_\tau \gamma_\tau\|_{L^2}}{\sqrt{2\pi c_1}} \sqrt{|a(\gamma_\tau)|}.$$

In particular, the function $\tau \mapsto a(\gamma_\tau)$ is strictly increasing. If $a(\gamma_\tau) > 0$ then

$$\frac{d}{d\tau}\sqrt{a(\gamma_\tau)} \geq c_3^{-1}\|\partial_\tau\gamma_\tau\|_{L^2}$$

and a similar inequality holds when $a(\gamma_\tau) < 0$. Integrating these from $\tau_0 = \log r + T$ to $\tau_1 = \log R - T$ (after splitting this interval into two according to the sign of $a(\gamma_\tau)$ if necessary) we obtain

$$\int_{\tau_0}^{\tau_1}\|\partial_\tau\gamma_\tau\|_{L^2}\,d\tau \leq c_3\left(\sqrt{|a(\gamma_{\tau_0})|} + \sqrt{|a(\gamma_{\tau_1})|}\right) \leq c_4\sqrt{\frac{E(u)}{T}}.$$

Since $\|\partial_\tau\gamma_\tau\|_{L^2} \leq c_5\sqrt{E(u)/T}$ for $\tau_0 \leq \tau \leq \tau_1$ this implies

$$E(u; A(e^T r, e^{-T}R)) = \int_{\tau_0}^{\tau_1}\|\partial_\tau\gamma_\tau\|_{L^2}^2\,d\tau \leq c_6\frac{E(u)}{T}.$$

Moreover,

$$
\begin{aligned}
\int_0^{2\pi}\mathrm{dist}(\gamma_{\tau_0}(\theta), \gamma_{\tau_1}(\theta))\,d\theta
&\leq \int_{\tau_0}^{\tau_1}\int_0^{2\pi}|\partial_\tau\gamma_\tau|\,d\theta d\tau \\
&\leq \sqrt{2\pi}\int_{\tau_0}^{\tau_1}\|\partial_\tau\gamma_\tau\|_{L^2}\,d\tau \\
&\leq c_7\sqrt{\frac{E(u)}{T}}
\end{aligned}
$$

and this proves the lemma. $\qquad\square$

It is convenient to introduce some notation. Let $u_\nu : \mathbb{C}P^1 \to M$ be any sequence of smooth maps. A point $z \in \mathbb{C}P^1$ is called **regular** for u_ν if there exists an $\varepsilon > 0$ such that the sequence du_ν is uniformly bounded on $B_\varepsilon(z)$. A point $z \in \Sigma$ is called **singular** for u_ν if it is not regular. This means that there exists a sequence $z_\nu \to z$ such that $|du_\nu(z_\nu)|$ is unbounded. A singular point z for u_ν is called **rigid** if it is singular for every subsequence of u_ν. This means that the sequence $z_\nu \to z$ can be chosen such that $|du_\nu(z_\nu)| \to \infty$.[2] It is called **tame** if it is isolated (no other singular points in some neighbourhood of z) and the limit

$$m_\varepsilon(z) = \lim_{\nu\to\infty}\int_{B_\varepsilon(z)}u_\nu^*\omega$$

exists and is finite for every sufficiently small $\varepsilon > 0$. In this case the **mass** of the singularity is defined to be the number

$$m(z) = \lim_{\varepsilon\to 0}m_\varepsilon(z).$$

This limit exists because the function $\varepsilon \mapsto m_\varepsilon(z)$ is non-decreasing. We shall prove that for a sequence of J-holomorphic curves each tame singularity has mass $m(z) \geq \hbar$ where $\hbar > 0$ is defined by the following lemma.

---

[2]**Warning:** If you identify $\mathbb{C}P^1$ with $\mathbb{C} \cup \{\infty\}$ then the point infinity is regular iff it is regular for the sequence $z \mapsto u_\nu(1/z)$. Similarly for singular and rigid. In particular, ∞ is a rigid singular point if there exists a sequence $z_\nu \to \infty$ such that $|du_\nu(z_\nu)|\cdot|z_\nu|^2 \to \infty$ where the norm of $du_\nu(z_\nu)$ is taken with respect to the standard metric on $\mathbb{C}$.

Lemma 4.5.2 *For every compact symplectic manifold (M,ω) and every almost complex structure $J \in \mathcal{J}_\tau(M,\omega)$ there exists a constant $\hbar > 0$ such that*

$$E(u) > \hbar$$

for every nonconstant J-holomorphic curve $u : \mathbb{C}P^1 \to M$.

Proof: Choose $\hbar > 0$ as in Lemma 4.2.2. Let $u : \mathbb{C} \cup \{\infty\} \to M$ be a J-holomorphic curve with $E(u) \leq \hbar$. Then for every $z \in \mathbb{C}$ and every $r > 0$ we have $|du(z)|^2 \leq 8E(u, B_r(z))/\pi r^2 \leq 8\hbar/\pi r^2$. Hence $du(z) = 0$ for every z and therefore u is constant. $\square$

We shall need the the following observation about complete metric spaces which is due to Hofer. The proof is left as an exercise.

Lemma 4.5.3 *Let X be a metric space and $f : X \to \mathbb{R}$ be continuous and nonnegative. Given $\zeta \in X$ and $\delta > 0$ assume that the closed ball $B_\delta(\zeta) = \{x \in X \mid \mathrm{dist}(x,\zeta) \leq \delta\}$ is complete. Then there exist $z \in X$ and $0 < \varepsilon < \delta$ such that*

$$\mathrm{dist}(z,\zeta) \leq \delta, \qquad \sup_{B_\varepsilon(z)} f \leq 2f(z), \qquad \varepsilon f(z) \geq \frac{\delta f(\zeta)}{2}.$$

Lemma 4.5.4 *Let J_ν be a sequence of almost complex structures on M converging to J in the C^∞-topology and let $u_\nu : \mathbb{C} \cup \{\infty\} \to M$ be a sequence of J_ν-holomorphic curves with bounded energy $\sup_\nu E(u_\nu) = c < \infty$. Then every rigid singular point z for u_ν has mass $m(z) \geq \hbar$.*

Proof: This is proved by a rescaling argument as in the proof of Theorem 4.3.2. Assume without loss of generality that 0 is a rigid singular point of u_ν. Then there exists a sequence $\zeta_\nu \to 0$ such that $|du_\nu(\zeta_\nu)| \to \infty$. Choose a sequence $\delta_\nu > 0$ such that

$$\delta_\nu \to 0, \qquad \delta_\nu |du_\nu(\zeta_\nu)| \to \infty.$$

Now use Lemma 4.5.3 with $X = \mathbb{C}$, $f = |du_\nu|$, $\zeta = \zeta_\nu$, and $\delta = \delta_\nu$ to obtain sequences $z_\nu \in \mathbb{C}$ and $0 < \varepsilon_\nu < \delta_\nu$ such that $z_\nu \to 0$ and

$$\sup_{B_{\varepsilon_\nu}(z_\nu)} |du_\nu| \leq 2|du_\nu(z_\nu)|, \qquad \varepsilon_\nu |du_\nu(z_\nu)| \to \infty.$$

Denote

$$c_\nu = |du_\nu(z_\nu)|, \qquad R_\nu = \varepsilon_\nu c_\nu$$

and consider the J_ν-holomorphic curves $v_\nu : B_{R_\nu} \to M$ defined by

$$v_\nu(z) = u_\nu(z_\nu + c_\nu^{-1}z).$$

These satisfy

$$\sup_{B_{R_\nu}} |dv_\nu| \leq 2, \quad |dv_\nu(0)| = 1, \quad R_\nu \to \infty, \quad E(v_\nu; B_{R_\nu}) \leq c.$$

Hence the sequence v_ν has a subsequence (still denoted by v_ν) which converges in the C^∞-topology to a J-holomorphic curve $v : \mathbb{C} \to M$. The energy of v is finite and hence it follows from the removable singularity theorem that v extends to a J-holomorphic sphere $v : \mathbb{C}P^1 = \mathbb{C} \cup \{\infty\} \to M$. Since $|dv(0)| = 1$ this sphere must be nonconstant and hence, by Lemma 4.5.2,

$$E(v) \geq \hbar.$$

Since $v(z) = \lim_{\nu \to \infty} u_\nu(z_\nu + c_\nu^{-1} z)$ we have

$$E(v; B_R) = \lim_{\nu \to \infty} E(v_\nu; B_{R/c_\nu}(z_\nu)) \leq \liminf_{\nu \to \infty} E(u_\nu; B_\varepsilon).$$

for every $R > 0$ (arbitrarily large) and every $\varepsilon > 0$ (arbitrarily small). The last inequality follows from the fact that R/c_ν converges to zero and z_ν converges to 0. Now take the limit $R \to \infty$ to obtain

$$\hbar \leq \liminf_{\nu \to \infty} E(u_\nu; B_\varepsilon).$$

Here we have chosen a subsequence of u_ν. But the same argument shows that every subsequence of u_ν has a further subsequence with this property and hence $m_\varepsilon(0) \geq \hbar$. Since $\varepsilon > 0$ was chosen arbitrarily this proves the lemma. $\square$

Lemma 4.5.5 *Let J_ν be a sequence of almost complex structures on M converging to J in the C^∞-topology and let $u_\nu : \mathbb{C} \cup \{\infty\} \to M$ be a sequence of J_ν-holomorphic curves with bounded energy*

$$\sup_\nu E(u_\nu) = E < \infty.$$

Then there exists a subsequence (still denoted by u_ν) which has only finitely many singular points $z^1, \ldots, z^k$ and these are all tame with positive mass $m(z^j) \geq \hbar$. The subsequence can be chosen to converge uniformly with all derivatives on every compact subset of $\mathbb{C}P^1 - \{z^1, \ldots, z^k\}$ to a J-holomorphic curve $u : \mathbb{C}P^1 \to M$ with energy

$$E(u) = E - \sum_{j=1}^{k} m(z^j).$$

Proof: By assumption there exists a constant $c > 0$ such that $E(u_\nu) \leq c$ for every ν. Hence it follows from Lemma 4.5.4 that the number of rigid singular points is bounded above by $c/\hbar$. Now choose a subsequence such that all its singular points are rigid. Such a subsequence must exist because if there is any nonrigid singular point left we may choose a further subsequence for which this singular point becomes rigid. This process must stop after finitely many steps since otherwise there would be a subsequence with arbitrarily many rigid singular points. By Lemma 4.5.4 this is impossible.

Now let $z^1, \ldots, z^k$ be the singular points of the subsequence u_ν and assume that they are all rigid. By definition of *singular point* the sequence of derivatives du_ν is uniformly bounded on every compact subset of $\mathbb{C}P^1 - \{z^1, \ldots, z^k\}$. Passing to a further subsequence we may assume that u_ν converges uniformly with all derivatives on every compact subset of $\Sigma - \{z^1, \ldots, z^k\}$. By the removable singularity theorem the limit extends to a J-holomorphic curve $u : \mathbb{C}P^1 \to M$.

Now fix a sufficiently small number $\varepsilon > 0$ and choose a further subsequence such that the limit $m_\varepsilon(z^j) = \lim_{\nu \to \infty} E(u_\nu; B_\varepsilon(z^j))$ exists for every j. Then this limit exists for every sufficiently small $\varepsilon > 0$ and hence the singular points z^j are all tame for u_ν. Now for $\varepsilon > 0$ sufficiently small

$$E(u; \mathbb{C} - \cup_j B_\varepsilon(z^j)) = \lim_{\nu \to \infty} E(u_\nu; \mathbb{C} - \cup_j B_\varepsilon(z^j)) = E - \sum_j m_\varepsilon(z^j)$$

Take the limit $\varepsilon \to 0$ to obtain the formula $E(u) = E - \sum_j m(z^j)$. $\square$

The phenomenon described in the previous lemma already occurs in the context of rational maps $u : \mathbb{C}P^1 \to \mathbb{C}P^1$. For example, the reader might like to consider a sequence of rational functions of degree 2 with a pole and a zero cancelling each other out in the limit.

Proof of Theorem 4.4.3: Passing to a suitably reparametrized subsequence (as in the proof of Theorem 4.3.2) we may assume that $u_\nu(1/z)$ converges uniformly on compact subsets of $\mathbb{C}$ to a nonconstant J-holomorphic curve $v^0 : \mathbb{C} \cup \{\infty\} \to M$. Hence $z = 0$ is the only possible singular point of the sequence u_ν and in view of Lemma 4.5.5 we may assume it is tame. Denote the mass of this singular point by

$$m_0 = \lim_{\varepsilon \to 0} m_\varepsilon, \qquad m_\varepsilon = \lim_{\nu \to \infty} \int_{B_\varepsilon} u_\nu^* \omega.$$

We will write u for the limit of u_ν on $\mathbb{C} - \{0\} \cup \infty$. Thus, $u(z) = v^0(1/z)$. Let us now examine the behaviour of the sequence u_ν near the singular point $z = 0$ in more detail. By Lemma 4.5.2 we have $m_0 \geq \hbar$ and hence for every ν there exists a number $\delta_\nu > 0$ such that

$$\int_{B_{\delta_\nu}} u_\nu^* \omega = m_0 - \frac{\hbar}{2}.$$

By definition of the mass m_0 the sequence δ_ν converges to 0. Consider the sequence of J-holomorphic maps $v_\nu : \mathbb{C} \to M$ defined by

$$v_\nu(z) = u_\nu(\delta_\nu z).$$

We shall prove that there exists a subsequence (still denoted by v_ν) such that the following holds.

(i) The singular set $\{w^1, \ldots, w^N\}$ of the subsequence v_ν is finite and tame and is contained in the open unit disc $B_1 \subset \mathbb{C}$.

(ii) The subsequence v_ν converges with all derivatives uniformly on every compact subset of $\mathbb{C} - \{w^1, \ldots, w^N\}$ to a nonconstant J-holomorphic curve $v : \mathbb{C} \to M$ with finite energy.

(iii) The energy of v and the masses of the singularities $w^1, \ldots, w^N$ are related by

$$E(v) + \sum_{j=1}^{N} m(w^j) = m_0.$$

(iv) $v(\infty) = u(0)$.

Once this is proved the theorem follows easily by induction. First note that by definition of m_0 and δ_ν

$$\limsup_{\nu \to \infty} E(v_\nu; A(1, R)) \leq \frac{\hbar}{2}$$

for every $R \geq 1$. Hence there can be no bubbling of holomorphic spheres outside the unit ball and so statements (i) and (ii) follow from Lemma 4.5.5. Now we shall prove that the limit curve $v : \mathbb{C} \to M$ satisfies

$$E(v; \mathbb{C} - B_1) = \frac{\hbar}{2}.$$

We already have proved that $E(v; \mathbb{C} - B_1) \leq \hbar/2$. To prove the converse choose a sequence $\varepsilon_\nu > 0$ such that $E(u_\nu; B_{\varepsilon_\nu}) = m_0$. Then it follows again from the definition of m_0 that $\varepsilon_\nu \to 0$. Now consider the sequence $w_\nu(z) = u_\nu(\varepsilon_\nu z)$. It follows as above that $E(w_\nu; A(1, R))$ converges to zero for any $R > 1$. This implies that $E(w_\nu; A(\delta, 1))$ must also converge to zero for any $\delta > 0$ since otherwise a subsequence of w_ν would converge to a nonconstant J-holomorphic curve which is constant for $|z| \geq 1$ but such a curve does not exist. Since

$$E(w_\nu; A(\delta_\nu/\varepsilon_\nu, 1)) = E(u_\nu; A(\delta_\nu, \varepsilon_\nu)) = \frac{\hbar}{2}$$

it follows that $\delta_\nu/\varepsilon_\nu$ converges to 0. Now, by Lemma 4.5.1, there exists a $T_0 > 0$ such that for $T > T_0$

$$E(u_\nu; A(e^T \delta_\nu, e^{-T}\varepsilon_\nu)) \leq \frac{c}{T} E(u_\nu; A(\delta_\nu, \varepsilon_\nu)) = \frac{c}{T}\frac{\hbar}{2}.$$

Pick any number $\alpha < 1$ and choose T so large that $1 - c/T > \alpha$. Then the energy of u_ν in the union of the annuli $A(\delta_\nu, e^T \delta_\nu)$ and $A(e^{-T}\varepsilon_\nu, \varepsilon_\nu)$ must be at least $\alpha\hbar/2$. But the energy of u_ν in $A(e^{-T}\varepsilon_\nu, \varepsilon_\nu)$ converges to 0 while the energy of u_ν in $A(\delta_\nu, e^T \delta_\nu)$ converges to $E(v; A(1, e^T))$. Hence $E(v; A(1, e^T)) \geq \alpha\hbar/2$. Since $\alpha < 1$ was chosen arbitrarily it follows that $E(v, \mathbb{C} - B_1) = \hbar/2$ as claimed.

Statement (iii) now follows from Lemma 4.5.5 for the curves $v_\nu : B_1 \to M$ with constant energy $E_{B_1}(v_\nu) = m_0 - \hbar/2$. Hence it remains to prove that $v(\infty) = u(0)$. We shall prove that the *collar* $u_\nu(z)$ for $R\delta_\nu < |z| < \varepsilon$ converges uniformly as $\varepsilon \to 0$, $R \to \infty$, and $\nu \geq \nu(\varepsilon, R) \to \infty$. To see this note that

$$E(\varepsilon, R) \stackrel{\text{def}}{=} \lim_{\nu \to \infty} E(u_\nu; A(R\delta_\nu, \varepsilon)) = E(u; B_\varepsilon) + E(v; \mathbb{C} - B_R).$$

In fact $E(u_\nu; B_\varepsilon)$ converges to $E(u; B_\varepsilon) + m_0$ as ν tends to ∞ and, by definition of δ_ν, it follows that $E(u_\nu; A(\delta_\nu, \varepsilon))$ converges to $E(u; B_\varepsilon) + \hbar/2$. On the other hand, $E(u_\nu; A(\delta_\nu, R\delta_\nu))$ converges to $E(v; A(1, R)) = \hbar/2 - E(v; \mathbb{C} - B_R)$. Subtracting these two limits gives the required formula above.

Now $E(\varepsilon, R)$ converges to zero as $\varepsilon \to 0$ and $R \to \infty$. Hence it follows from Lemma 4.5.1 that for $T > 0$ and $\nu > 0$ sufficiently large we have

$$\int_0^{2\pi} \text{dist}(u_\nu(R\delta_\nu e^{T+i\theta}), u_\nu(\varepsilon e^{-T+i\theta}))\, d\theta \leq c\sqrt{E(\varepsilon, R)}.$$

Taking the limit $\nu \to \infty$ we obtain

$$\int_0^{2\pi} \mathrm{dist}(v(Re^{T+i\theta}), u(\varepsilon e^{-T+i\theta}))\, d\theta \le c\sqrt{E(\varepsilon, R)}.$$

Here the constants T and c are independent of ε and R. Since $E(\varepsilon, R) \to 0$ for $\varepsilon \to 0$ and $R \to \infty$ we obtain $v(\infty) = u(0)$ as required. Theorem 4.4.3 now follows easily by induction. $\qquad\square$

Chapter 5

Compactification of Moduli Spaces

Our goal in this chapter is to explain the compactification $\overline{\mathcal{C}}(A, J)$ of the moduli space $\mathcal{C}(A, J) = \mathcal{M}(A, J)/G$ of all J-holomorphic spheres which represent the class A. Roughly speaking, this compactification is obtained by adding the cusp-curves to the space $\mathcal{C}(A, J)$. In order for this compactification to be useful, it should carry a fundamental homology class. This will be the case if the set of cusp-curves which we must add has dimension at least 2 less than that of $\mathcal{C}$. Unfortunately, it is not known whether this is true for an arbitrary symplectic manifold. However, we will see that it does hold for manifolds which satisfy a certain positivity (or monotonicity) condition. This is an important condition, and so we will begin by explaining it.

Our present approach modifies and streamlines the discussion in McDuff [46] and Ruan [64]. The main results are the Theorems 5.2.1, 5.3.1 and 5.4.1. Their proofs are deferred until Chapter 6.

5.1   Semi-positivity

Let (M, ω) be a symplectic manifold and let $K > 0$. An ω-compatible almost complex structure $J \in \mathcal{J}(M, \omega)$ is called K**-semi-positive** if every J-holomorphic sphere $u : \mathbb{C}P^1 \to M$ with energy $E(u) \leq K$ has nonnegative Chern number $\int u^* c_1 \geq 0$.[1] The almost complex structure J is called **semi-positive** if it is K-semi-positive for every K. This means that every J-holomorphic curve $u : \mathbb{C}P^1 \to M$ has nonnegative Chern number. The notions of K**-positive** and **positive** are defined similarly with $\int u^* c_1 > 0$. Denote by $\mathcal{J}_+(M, \omega, K)$ the set of all K-semi-positive almost complex structures on M which are ω-compatible and by

$$\mathcal{J}_+(M, \omega) = \bigcap_{K > 0} \mathcal{J}_+(M, \omega, K)$$

the set of semi-positive ω-compatible almost complex structures. Note that this set may be empty. The manifold (M, ω) is called **weakly monotone** if for every

---

[1] Here we are identifying the first Chern class c_1 with a representing 2-form.

spherical homology class $A \in H_2(M, \mathbb{Z})$

$$\omega(A) > 0, \quad c_1(A) \geq 3 - n \quad \Longrightarrow \quad c_1(A) \geq 0.$$

It is called **monotone** if there exists a number $\lambda > 0$ such that

$$\omega(A) = \lambda c_1(A)$$

for $A \in \pi_2(M)$. Note that every monotone symplectic manifold is weakly monotone. A complex manifold (M, J) (of complex dimension n) is called a **Fano variety** if its anti-canonical bundle $K^* = \Lambda^{0,n} T^* M$ is ample. This is equivalent to the existence of a Kähler metric such that the Kähler form ω is *monotone*.

Remark 5.1.1 In Chapters 5 and 6 we shall work with the space $\mathcal{J}(M, \omega)$ of almost complex structures which are ω-compatible. However, all the transversality and compactness theorems remain valid if we take interpret $\mathcal{J}(M, \omega)$ as the space of all ω-tame almost complex structures, replacing the word ω-compatible by ω-tame wherever it occurs.

Lemma 5.1.2 *Let M be a compact $2n$-dimensional manifold. Then for all $K > 0$ the set*

$$\left\{ (\omega, J) \in \Omega^2(M) \times \mathcal{J}(M, \omega) \,|\, d\omega = 0, \ \omega^n \neq 0, \ J \in \mathcal{J}_+(M, \omega, K) \right\}$$

is open in the space of all compatible pairs (ω, J) with respect to the C^1-topology. In particular, for every symplectic form ω the set $\mathcal{J}_+(M, \omega, K)$ is open in $\mathcal{J}(M, \omega)$ with respect to the C^1-topology.

Proof: Let ω_ν be a sequence of symplectic forms on M which converge to a symplectic form ω in the C^1-topology. Let $J_\nu \in \mathcal{J}(M, \omega_\nu)$ be a sequence of almost complex structures on M which are not K-semi-positive with respect to ω_ν. Assume that J_ν converges to $J \in \mathcal{J}(M, \omega)$ in the C^1-topology. Then there exists a sequence $u_\nu : \mathbb{C}P^1 \to M$ of J_ν-holomorphic curves with $c_1(u_\nu) < 0$ and $E(u_\nu) \leq K$. By Theorem 4.4.3 u_ν has a subsequence which converges weakly (in the $W^{2,p}$-topology) to a cusp-curve $u = (u^1, \dots, u^k)$. One of the curves u^j must have negative Chern number and they all have energy $E(u^j) \leq K$. Hence J is not K-semi-positive with respect to ω. $\qquad\square$

In general, the set of semi-positive almost complex structures on a compact symplectic manifold will not be open. If J is semi-positive then the size of a neighbourhood which consists of K-semi-positive structures may depend on K. However, the next lemma shows that if the manifold (M, ω) is weakly monotone then the set $\mathcal{J}_+(M, \omega)$ is dense and path-connected.

Lemma 5.1.3 *Assume that (M, ω) is a weakly monotone compact symplectic manifold. Then the set $\mathcal{J}_+(M, \omega)$ contains a path-connected dense subset. The set $\mathcal{J}_+(M, \omega, K)$ is open, dense, and path connected for every K.*

Proof: We first prove that every almost complex structure $J \in \mathcal{J}_{\mathrm{reg}}(M, \omega)$ which is regular in the sense of Definition 3.1.1 is semi-positive. To see this let $J \in \mathcal{J}_{\mathrm{reg}}(M, \omega)$ and $A \in H_2(M, Z)$ with $c_1(A) < 0$. We must prove that $\mathcal{M}(A, J) = \emptyset$.

Assume otherwise that $\mathcal{M}(A, J) \neq \emptyset$. Since every curve $u \in \mathcal{M}(A, J)$ is simple we must have $\omega(A) > 0$ and, since (M, ω) is weakly monotone this implies

$$c_1(A) \leq 2 - n.$$

But the moduli space $\mathcal{M}(A, J)/G$ has dimension

$$\dim \mathcal{M}(A, J)/G = 2n + 2c_1(A) - 6$$

and the above condition on $c_1(A)$ shows that this number is negative. This contradicts our assumption that $\mathcal{M}(A, J)$ be nonempty. Thus we have proved that $\mathcal{J}_{\text{reg}}(M, \omega) \subset \mathcal{J}_+(M, \omega)$. Now the set $\mathcal{J}_{\text{reg}}(M, \omega)$ need not be connected. However, Theorem 3.1.3 shows that any two points $J_0, J_1 \in \mathcal{J}_{\text{reg}}(M, \omega)$ can be connected by a path $[0, 1] \to \mathcal{J}(M, \omega) : \lambda \mapsto J_\lambda$ such that the quotient space $\mathcal{M}(A, \{J_\lambda\}_\lambda)/G$ is a manifold of dimension

$$\dim \mathcal{M}(A, \{J_\lambda\}_\lambda)/G = 2n + 2c_1(A) - 5.$$

If $c_1(A) < 0$ it follows again that this dimension is negative and so for every λ the space $\mathcal{M}(A, J_\lambda)$ must be empty. Hence $J_\lambda \in \mathcal{J}_+(M, \omega)$ for every λ. This proves the lemma. $\qquad\square$

The condition that (M, ω) be weakly monotone depends only on the homotopy classes of ω and $c_1(M)$, and is the relevant positivity condition to consider in the case of a symplectic manifold. If one starts with a Kähler manifold though, one is often more interested in the properties of the underlying complex manifold (M, J_0) than in the cohomology class of the chosen Kähler form ω. We will see in Proposition 7.2.3 below that our results apply to complex manifolds (M, J_0) such that J_0 is compatible with some weakly monotone Kähler (or symplectic) form ω.

Remark 5.1.4 It is not hard to check that a symplectic manifold (M, ω) is weakly monotone if and only if one of the following conditions is satisfied.

- (M, ω) is monotone.
- either $c_1(A) = 0$ for every $A \in \pi_2(M)$, or $\omega(A) = 0$ for every $A \in \pi_2(M)$.
- The minimal Chern number N, defined by $\langle c_1, \pi_2(M) \rangle = N\mathbb{Z}$ where $N \geq 0$, is at least $N \geq n - 2$.

These conditions are not mutually exclusive. The last condition shows that every symplectic manifold of dimension less than or equal to 6 is weakly monotone. The case when $\omega(A)$ is always zero is not very interesting in the present context, since obviously there cannot be any J-holomorphic spheres in such manifolds. However, there could be J-holomorphic curves of higher genus. $\qquad\square$

As an example, let M be a product of a number of projective spaces, $M = \mathbb{C}P^{n_1} \times \ldots \times \mathbb{C}P^{n_k}$, with form $\sum_i a_i \tau_i$, where τ_i is the usual Kähler form, normalised so that it integrates to π over each complex line. Then M is monotone only if all the a_i are equal. However, it is easy to see that the standard complex structure J_0 is semi-positive. Hence, for each $K > 0$ there exists a neighbourhood of J_0 which consists entirely of K-semi-positive almost complex structures.

5.2   The image of the evaluation map

We now describe a first result concerning the compactification of the moduli space. There are different ways in which this result can be stated. Instead of specifying a topology on the closure $\overline{C}$ of the space of unparametrized curves, we have chosen to express the fact that $\overline{C}$ is closed by stating that it has closed image in M, since this is what is needed for the applications we have in mind. Thus we consider the image of our spaces of curves by appropriate evaluation maps. We will use the letter $\mathcal{W}$ to describe those moduli spaces which are the natural domains of definition for evaluation maps.

Recall from Chapter 3 that the reparametrization group $G = PSL(2, \mathbb{C})$ acts on $\mathcal{M}(A, J) \times \mathbb{C}P^1$ by

$$\phi \cdot (u, z) = (u \circ \phi^{-1}, \phi(z)).$$

The quotient space

$$\mathcal{W}(A, J) = \mathcal{M}(A, J) \times_G \mathbb{C}P^1$$

is the domain of the evaluation map

$$e : \mathcal{W}(A, J) \to M, \qquad e(u, z) = u(z).$$

We shall denote the image of this map by

$$X(A, J) = e(\mathcal{W}(A, J)) = \bigcup_{u \in \mathcal{M}(A, J)} u(\mathbb{C}P^1).$$

Our goal is to describe the closure $\overline{X}(A, J)$ as a kind of stratified space with open stratum $X(A, J)$ and all other strata of codimension at least 2. We must prove that the lower dimensional cover all possible limit points of sequences $e(u_\nu, z_\nu)$ where the sequence $[u_\nu, z_\nu]$ does not converge in $\mathcal{W}(A, J)$. These other strata are defined, roughly speaking, as the images of evaluation maps on spaces of *simple cusp-curves* which represent the class A. A cusp-curve

$$C = C^1 \cup \ldots \cup C^N$$

is called **simple**[2] if all its components C^i are distinct and none are multiply-covered. More precisely, a simple cusp-curve is a collection of simple J-holomorphic spheres

$$u^i : \mathbb{C}P^1 \to M$$

such that their images form a connected set and $u^i \neq u^j \circ \phi$ for $i \neq j$ and any fractional linear transformation $\phi \in G$. Observe that every cusp-curve can be simplified to a simple curve, by getting rid of repeated components and replacing each multiply covered component by its underlying simple component. Clearly, this process does not change the set of points which lie on the curve, although it will change its homology class.

---

[2]In [46] and [64] such curves were called reduced.

Theorem 5.2.1 *Let (M, ω) be a compact symplectic manifold.*

(i) *For every ω-compatible almost complex structure $J \in \mathcal{J}(M, \omega)$ there exists a finite collection of evaluation maps $e_D : \mathcal{W}(D, J) \to M$ such that*

$$\bigcap_{\substack{K \subset \mathcal{W}(A, J) \\ K \text{ compact}}} \overline{e(\mathcal{W}(A, J) - K)} \subset \bigcup_D e_D(\mathcal{W}(D, J)).$$

(ii) *There exists a set $\mathcal{J}_{\mathrm{reg}} = \mathcal{J}_{\mathrm{reg}}(M, \omega, A, 1) \subset \mathcal{J}(M, \omega)$ of the second category such that the set $\mathcal{W}(D, J)$ is a smooth oriented σ-compact manifold of dimension*

$$\dim \mathcal{W}(D, J) = 2n + 2c_1(D) - 2N - 2$$

for $J \in \mathcal{J}_{\mathrm{reg}}$. Here N is the number of components of the class D.

(iii) *Assume that A is not a multiple class mB where $m > 1$ and $c_1(B) = 0$. If $J \in \mathcal{J}_+(M, \omega, K) \cap \mathcal{J}_{\mathrm{reg}}$ for some $K > \omega(A)$ then*

$$\dim \mathcal{W}(D, J) \leq \dim \mathcal{W}(A, J) - 2$$

for every $D \neq A$.

In the terminology of Section 7.1, the above theorem asserts that the evaluation map $e = e_J : \mathcal{W}(A, J) \to M$ determines a *pseudo-cycle* for a generic almost complex structure J. This means that its image can be compactified by adding pieces of codimension at least 2, and hence that it carries a fundamental class. We shall see below that this homology class is independent of the almost complex structure J used to define it. (See Proposition 7.2.2 with $p = 1$.) The proof involves choosing a regular path of almost complex structures J_λ running from J_0 to J_1 such that the corresponding space $\mathcal{W}(D, \{J_\lambda\}_\lambda)$ is a smooth manifold. This manifold forms a cobordism from $\mathcal{W}(D, J_0)$ to $\mathcal{W}(D, J_1)$ and there is an analogue of Theorem 5.2.1 for the evaluation map on $\mathcal{W}(D, \{J_\lambda\}_\lambda)$. In short, the evaluation maps e_{J_0} and e_{J_1} are bordant as pseudo-cycles.

The framing D

We now give the relevant definitions for the notation in Theorem 5.2.1. The letter D denotes an effective framed class. This means that

$$D = \{A^1, \ldots, A^N, j_2, \ldots, j_N\}$$

where $A^i \in H_2(M, \mathbb{Z})$ and j_ν is an integer with $1 \leq j_\nu \leq \nu - 1$. The homology classes A^i are effective in the sense that they each have J-holomorphic representatives and can be the classes of the components of a simplified cusp-curve C which represents the class A. For example there is a necessary condition $\omega(A^i) \leq \omega(A)$ for all i and hence it follows from Corollary 4.4.4 that there is only a finite number of possible framed classes D. The "framing" of D describes how the different components intersect. The number j_ν indicates that the cusp-curve $u = (u^1, \ldots, u^N)$ must satisfy

$$u^\nu(\mathbb{C}P^1) \cap u^{j_\nu}(\mathbb{C}P^1) \neq \emptyset.$$

Note that the class D of a cusp-curve can never equal $\{A\}$ itself since either $N > 1$ or $D = \{A^1\}$ where $A = mA^1$ for some $m > 1$.

We will see in Chapter 6 that, in order to make $\mathcal{W}(D, J)$ into a manifold, we must restrict to considering simple cusp-curves. In fact, we saw already in Chapter 3 why we must avoid multiply covered curves, and repeated components can also cause trouble. This means that $\sum_i A^i$ need not equal A, since the process of simplification changes the homology class. However, there are integers $m_i \geq 1$ such that

$$\sum_i m_i A^i = A.$$

If J is K-semi-positive then $c_1(A^i) \geq 0$ for all i and so

$$c_1(D) = \sum_i c_1(A^i) \leq c_1(A).$$

This is the basic reason why statement (iii) in Theorem 5.2.1 holds in this case. Observe also that the set of points which lie on C is unchanged if we replace C by its simple reduction. Hence restricting to simple C does not affect the validity of (i).

We remark that the framing (or intersection pattern) associated to D describes enough of the intersection pattern of the different components of C to ensure that C is connected. For example, one might insist that the A^2-curve intersects the A^1-curve, and that the A^3 curve also intersects the A^1-curve, and so on. Because we do not insist on describing the full intersection pattern, our approach is somewhat simpler than that in [64]. Also, there are cusp-curves which belong to more than one set $\mathcal{C}(D, J)$. Further details may be found in Chapter 6. Elements of $\mathcal{C}(D, J)$ are called simple cusp-curves of type D.

The spaces $\mathcal{W}(D, J)$

According to our conventions, the spaces $\mathcal{W}(D, J)$ are the correct spaces on which evaluation maps are defined. A precise definition will be given in Chapter 6. As an example consider the case $D = \{A, B\}$ with $A \neq B$. In this case there is only one possible intersection pattern, namely the A-curve must intersect the B-curve. Fix a point $z_0 \in \mathbb{C}P^1$ and define $G_0 = \{\phi \in G \mid \phi(z_0) = z_0\}$. The group $G_0 \times G_0$ acts on the space

$$\mathcal{M}(D, J) = \{(u, v) \mid u \in \mathcal{M}(A, J),\, v \in \mathcal{M}(B, J),\, u(z_0) = v(z_0)\}.$$

This space is the inverse image of the diagonal $\Delta \subset M \times M$ under the map $\mathcal{M}(A, J) \times \mathcal{M}(B, J) : (u, v) \mapsto (u(z_0), v(z_0))$ and hence, by a general position argument with generic J, will have dimension

$$\dim \mathcal{M}(D, J) = 2n + 2c_1(A).$$

Now the group $G_0 \times G_0$ acts on the space $\mathcal{M}(D, J) \times \mathbb{C}P^1$ in two ways, namely by $(\phi_1, \phi_2) \cdot (u, v, z) = (u \circ \phi_1^{-1}, v \circ \phi_2^{-1}, \phi_j(z))$ for $j = 1, 2$. This leads to two different quotient spaces $\mathcal{W}_1(D, J)$ and $\mathcal{W}_2(D, J)$ with corresponding evaluation maps $e_A, e_B : \mathcal{W}_j(D, J) \to M$ defined by

$$e_A(u, v, z) = u(z), \quad e_B(u, v, z) = v(z).$$

We define $\mathcal{W}(D,J) = \mathcal{W}_1(D,J) \cup \mathcal{W}_2(D,J)$. Since the group $G_0 \times G_0$ is 8-dimensional and in both cases acts freely on $\mathcal{M}(D,J) \times \mathbb{C}P^1$ we obtain

$$\dim \mathcal{W}(D,J) = 2n + 2c_1(A) - 6.$$

This is precisely 2 less than the dimension of $\mathcal{W}(A,J)$, thus confirming the dimension formula in statement (ii) of Theorem 5.2.1.

Granted these definitions, part (i) of Theorem 5.2.1 follows immediately from Gromov's compactness theorem. The main additional technical point, which one needs in order to control the dimension of the spaces $\mathcal{W}(D,J)$, is a proof that for generic J the evaluation maps

$$e_A : \mathcal{M}(A,J) \times S^2 \to M, \qquad e_B : \mathcal{M}(B,J) \times S^2 \to M$$

are transverse. It is here that the condition $u \neq v \circ \phi$ becomes important. The details will be carried out in Chapter 6.

5.3 The image of the p-fold evaluation map

In order to define the Gromov-Witten invariants we must also compactify the image $X(A,J,p)$ of the p-fold evaluation map

$$e_p : \mathcal{W}(A,J,p) \to M^p$$

defined on the quotient space $\mathcal{W}(A,J,p) = \mathcal{M}(A,J) \times_G (\mathbb{C}P^1)^p$. Here the group $G = \mathrm{PSL}(2,\mathbb{C})$ acts on $\mathcal{M}(A,J) \times (\mathbb{C}P^1)^p$ by

$$\phi \cdot (u, z_1, \ldots, z_p) = (u \circ \phi^{-1}, \phi(z_1), \ldots, \phi(z_p))$$

and the evaluation map is given by

$$e_p(u, z_1, \ldots, z_p) = (u(z_1), \ldots, u(z_p)).$$

Denote the image of this evaluation map by $X(A,J,p) = e_p(\mathcal{W}(A,J,p)) \subset M^p$. Essentially the same result applies in this case. Suppose that the A-curves $u_\nu : \mathbb{C}P^1 \to M$ converge to a simple cusp-curve C of type

$$D = \{A^1, \ldots, A^N, j_2, \ldots, j_N\}$$

and consider possible limit points of the p-tuples $(u_\nu(z_{1\nu}), \ldots, u_\nu(z_{p\nu}))$. After passing to a subsequence, if necessary, the the points $u_\nu(z_{j\nu})$ will converge to a point z_j on some component of C. Hence we must consider all possible configurations of limit points $z_j \in C_{T(j)}$. Thus the domain of our p-fold evaluation maps will be determined by pairs (D,T) where D is a framed class of simple reduced curves as before and the label T is a map from $\{1, \ldots, p\} \to \{1, \ldots, N\}$. For each such pair (D,T) there will be an evaluation map

$$e_{D,T} : \mathcal{W}(D,T,J,p) \to M^p$$

whose image $X(D,T,J,p) = e_{D,T}(\mathcal{W}(D,T,J,p)) \subset M^p$ consists of all p-tuples of points $(x_1, \ldots, x_p)$ such that $x_j \in C_{T(j)}$ where $C = \cup_i C_i$ runs through all simple cusp-curves of type D. These images will form the closure of the set $X(A,J,p)$.

Theorem 5.3.1 *Let (M, ω) be a compact symplectic manifold.*

(i) *For every $J \in \mathcal{J}(M, \omega)$ there exists a finite collection of evaluation maps $e_{D,T}$: $\mathcal{W}(D, T, J, p) \to M^p$ such that*

$$\bigcap_{\substack{K \subset \mathcal{W}(A, J, p) \\ K \ compact}} \overline{e_p(\mathcal{W}(A, J, p) - K)} \subset \bigcup_{D, T} e_{D,T}(\mathcal{W}(D, T, J, p)).$$

(ii) *There exists a set $\mathcal{J}_{\mathrm{reg}} = \mathcal{J}_{\mathrm{reg}}(M, \omega, A, p) \subset \mathcal{J}(M, \omega)$ of the second category such that the set $\mathcal{W}(D, T, J, p)$ is a smooth oriented σ-compact manifold of dimension*

$$\dim \mathcal{W}(D, T, J, p) = 2n + 2c_1(D) + 2p - 2N - 4$$

for $J \in \mathcal{J}_{\mathrm{reg}}$. Here N is the number of components of the class D.

(iii) *Assume that A is not a multiple class mB where $m > 1$ and $c_1(B) = 0$. If $J \in \mathcal{J}_+(M, \omega, K) \cap \mathcal{J}_{\mathrm{reg}}$ for some $K > \omega(A)$ then*

$$\dim \mathcal{W}(D, T, J, p) \leq \dim \mathcal{W}(A, J, p) - 2$$

for all $D \neq A$.

This may be proved by the same methods as the previous theorem. The main problem is in correctly defining the sets $\mathcal{W}(D, T, J, p)$ and this will be done in the Section 6.5. Again, this theorem asserts that for a generic almost complex structure J the evaluation map $e_p : \mathcal{W}(A, J, p) \to M^p$ determines a *pseudo-cycle* in the sense of Section 7.1. Hence its image carries a fundamental homology class, and we shall see that this class is independent of the choice of J (see Proposition 7.2.2). This suffices to define the Gromov invariant Φ.

5.4 The evaluation map for marked curves

There is an alternative way of defining a p-fold evaluation map, namely by fixing a p-tuple

$$\mathbf{z} = (z_1, \ldots, z_p) \in (\mathbb{C}P^1)^p$$

of distinct points in $\mathbb{C}P^1$ and evaluating a (parametrized) J-holomorphic curve u at the points z_j. Hence define the evaluation map

$$e_{\mathbf{z}} = e_{A, J, \mathbf{z}} : \mathcal{M}(A, J) \to M^p$$

by

$$e_{\mathbf{z}}(u) = (u(z_1), \ldots, u(z_p)).$$

This map is used to define the Gromov-Witten invariant Ψ (which Ruan called $\tilde{\Phi}$) and will play an important role in the definition of quantum cohomology. The natural assumption for the definition of this invariant is $p \geq 3$ since three points on a nonconstant curve determine its parametrization. In fact in the case $p = 3$ we shall see that both invariants agree and no new construction is required. However, for $p > 3$ there are genuine differences. This will already be apparent in the proof of the following compactness theorem for the images

$$Y(A, J, \mathbf{z}) = e_{\mathbf{z}}(\mathcal{M}(A, J)) \subset M^p.$$

It is worth pointing out that for any triple $\mathbf{z} = (z_1, z_2, z_3)$ of distinct points in $\mathbb{C}P^1$ we have $Y(A, J, \mathbf{z}) \subset X(A, J, 3)$ and the closures of both sets agree. The sets themselves need not be the same since in the definition of X the points on which u is evaluated are not required to be distinct.

For $p > 3$ there is still an inclusion $Y(A, J, \mathbf{z}) \subset X(A, J, p)$. However the set $Y(A, J, \mathbf{z})$ will have strictly lower dimension. To get a picture of how these spaces fit together for different choices of $\mathbf{z}$, consider the diagram

$$\begin{array}{ccc} \mathcal{M}(A, J) \times (\mathbb{C}P^1)^p & \overset{\tilde{e}_p}{\to} & M^p \\ \pi \downarrow & & \\ (\mathbb{C}P^1)^p & & \end{array}$$

The set $Y(A, J, \mathbf{z})$ is simply the image of the fiber $\pi^{-1}(\mathbf{z})$ under the obvious evaluation map $\tilde{e}_p$. Thus the sets $Y(A, J, \mathbf{z})$ may be thought of as forming the fibers of a (singular) fibration of $X(A, J, p)$ of codimension $2p - 6$.

As before our goal is to prove that the boundary of the set $Y(A, J, \mathbf{z})$ is a finite union of lower dimensional strata. Thus we must investigate how the images of the manifolds $\mathcal{W}(D, T, J, p)$ which compactify $X(A, J, p)$ intersect the fibers $Y(A, J, \mathbf{z})$. In particular, in order to give an accurate description of the possible limiting positions of the images of the points z_i, we must keep track of the points at which the bubbles occur. For example, if u_ν is a sequence in $\mathcal{M}(A, J)$ which develops a bubble at z_1, then the sequence of points $u_\nu(z_1)$ may have subsequences which converge to any point of the bubble at z_1. It follows that the whole bubble is contained in $\overline{Y}(A, J, \mathbf{z})$. Multiply-covered components of the limit also pose new problems. The upshot is that we can only prove that a good compactification exists under additional assumptions.

More precisely, we will assume that our almost complex structure $J \in \mathcal{J}(M, \omega)$ and the homology class A satisfy the following condition.

(JA_p) Every J-effective homology class $B \in H_2(M, \mathbb{Z})$ has Chern number

$$c_1(B) \geq 2.$$

Moreover, if $A = mB \in H_2(M, \mathbb{Z})$ is the m-fold multiple of a J-effective homology class $B \in H_2(M, \mathbb{Z})$ with $m > 1$ then either $c_1(B) \geq 3$ or $p \leq 2m$.

Here a homology class $A \in H_2(M, \mathbb{Z})$ is called **J-effective** if it can be represented by a J-holomorphic sphere $u : \mathbb{C}P^1 \to M$. In particular a J-effective homology class is necessarily spherical. These assumptions are stronger than what is actually needed. For example in the case $p = 3$ it suffices to assume

(JA_3) Every J-effective homology class $B \in H_2(M, \mathbb{Z})$ has Chern number

$$c_1(B) \geq 0.$$

Moreover, if $A = mB \in H_2(M, \mathbb{Z})$ is the m-fold multiple of a J-effective homology class $B \in H_2(M, \mathbb{Z})$ with $m > 1$ then $c_1(B) \geq 1$.

This is precisely the condition of Theorem 5.3.1 (iii). In the case $p = 4$ only the following condition is required.

(JA_4) Every J-effective homology class $B \in H_2(M, \mathbb{Z})$ has Chern number

$$c_1(B) \geq 1.$$

Moreover, if $A = mB \in H_2(M, \mathbb{Z})$ is the m-fold multiple of a J-effective homology class $B \in H_2(M, \mathbb{Z})$ with $m > 1$ then $c_1(B) \geq 2$.

For general p the above condition can also be weakened but the precise condition required with the methods of this book is somewhat complicated to state. So we content ourselves with the above formulation.

We point out that for the definition of quantum cohomology it suffices to consider the cases $p = 3$ and $p = 4$. So in this case we may simply assume (JA_3) and (JA_4) for all classes A. These conditions will be satisfied, for example, if (M, ω) is monotone with minimal Chern number at least 2.

Now it would be much nicer, of course, if statement (iii) of the following theorem could be proved for generic J without any further assumptions on J. But this would require a better understanding of the behaviour of multiply covered J-holomorphic curves with Chern number less than or equal to 2 and with the present techniques it is not clear how to do this.

Theorem 5.4.1 *Let (M, ω) be a compact symplectic manifold and fix a homology class $A \in H_2(M)$ and a p-tuple $\mathbf{z} = (z_1, \ldots, z_p) \in (\mathbb{C}P^1)^p$ of distinct points in $\mathbb{C}P^1$.*

(i) *For every $J \in \mathcal{J}(M, \omega)$ there exists a finite collection of evaluation maps*

$$e_{D,T,\mathbf{z}} : \mathcal{V}(D, T, J, \mathbf{z}) \to M^p$$

such that

$$\bigcap_{\substack{K \subset \mathcal{M}(A,J) \\ K \text{ compact}}} \overline{e_{\mathbf{z}}(\mathcal{M}(A, J) - K)} \subset \bigcup_{D,T} e_{D,T,\mathbf{z}}(\mathcal{V}(D, T, J, \mathbf{z})).$$

(ii) *There exists a set $\mathcal{J}_{\mathrm{reg}} = \mathcal{J}_{\mathrm{reg}}(M, \omega, A, \mathbf{z}) \subset \mathcal{J}(M, \omega)$ of the second category such that the set $\mathcal{V}(D, T, J, \mathbf{z})$ is a finite dimensional smooth oriented σ-compact manifold for every $J \in \mathcal{J}_{\mathrm{reg}}$.*

(iii) *Assume $J \in \mathcal{J}_{\mathrm{reg}}$ and the pair (J, A) satisfies (JA_p). Then*

$$\dim \mathcal{V}(D, T, J, \mathbf{z}) \leq \dim \mathcal{M}(A, J) - 2$$

for all $D \neq A$.

This theorem is proved in Section 6.6. As before, this theorem asserts that for a generic almost complex structure J the evaluation map $e_{\mathbf{z}} : \mathcal{M}(A, J) \to M^p$ determines a *pseudo-cycle* in the sense of Section 7.1 and so its image carries a fundamental homology class. We shall see below that this class is independent of the point $\mathbf{z} \in (\mathbb{C}P^1)^p$ and of the almost complex structure J used to define it (Lemma 7.4.1). The proof of this fact again relies on the construction of a bordism from $e_{\mathbf{z}_0} : \mathcal{M}(A, J_0) \to M^p$ to $e_{\mathbf{z}_1} : \mathcal{M}(A, J_1) \to M^p$. This requires an analogue of Theorem 5.4.1 for paths along the following lines. Given a path $[0, 1] \to (\mathbb{C}P^1)^p : \lambda \mapsto \mathbf{z}_\lambda$ and regular almost complex structures $J_0 \in \mathcal{J}_{\mathrm{reg}}(M, \omega, A, \mathbf{z}_0)$ and $J_1 \in \mathcal{J}_{\mathrm{reg}}(M, \omega, A, \mathbf{z}_1)$ in the sense of Theorem 5.4.1, we must find a path of almost complex structures $J_\lambda \in \mathcal{J}(M, \omega)$ such that the moduli space

$$\mathcal{M}(A, \{J_\lambda\}_\lambda) = \{(\lambda, u) \,|\, u \in \mathcal{M}(A, J_\lambda)\}$$

is a smooth manifold. Now consider the closure of the image of the evaluation map

$$\mathcal{M}(A, \{J_\lambda\}_\lambda) \to M^p : (\lambda, u) \mapsto e_{\mathbf{z}_\lambda}(u).$$

For a generic path J_λ this closure will again be obtained by adding pieces of codimension at least 2. The details are the same as in the proof of Theorem 5.4.1 and this gives rise to the required bordism between the pseudo-cycles $e_{\mathbf{z}_0} : \mathcal{M}(A, J_0) \to M^p$ and $e_{\mathbf{z}_1} : \mathcal{M}(A, J_1) \to M^p$. This suffices to construct the Gromov-Witten invariant Ψ.

Remark 5.4.2 (i) The strong condition (JA_p) is used only to deal with multiply-covered components of the limiting cusp-curve. If there is some way to ensure that such multiply-covered curves cannot appear in the limit, then it suffices to assume that the manifold is weakly monotone. We shall exploit this observation in Section 9.3.

(ii) We emphasize that the notion of a regular almost complex structure in Theorem 5.4.1 depends on the point $\mathbf{z}$ which we have fixed to begin with. In practice it may be useful to fix instead an almost complex structure J which is regular in the sense of Theorem 5.3.1, and then vary the point $\mathbf{z}$ to achieve transversality.

$$\square$$

Chapter 6

Evaluation Maps and Transversality

This chapter contains the proofs of the main results of Chapter 5 and the reader may wish to go directly to Chapter 7. As we explained in Chapter 5 the pieces which must be added to compactify $\mathcal{M}(A, J)$ correspond to the cusp-curves. Our first task is to show that these cusp-curves appear in finite-dimensional families, and to calculate their dimension. This is accomplished in the first four sections. The last two sections contain the proofs of the compactness theorems of Chapter 5.

6.1   Evaluation maps are submersions

Every embedded symplectic 2-sphere is a J-holomorphic curve for some almost complex structure J which is compatible with ω. Since the symplectic condition is open it follows that the set of all points which lie on embedded J-holomorphic curves is open. This suggests that the evaluation map $e_0 : \mathcal{M}(A, \mathcal{J}) \to M$ at the point $z_0 \in \mathbb{C}P^1$ should be a submersion. Our goal in this section is to prove this for all simple curves, not just the embedded ones. Here we denote by $\mathcal{J} = \mathcal{J}(M, \omega)$ the space of almost complex structures on M which are compatible with ω. We could equally well consider the space of all ω-tame J: the only difference would be in the formula for the tangent space $T_J \mathcal{J}$. The evaluation map

$$e_0 : \mathcal{M}(A, \mathcal{J}) \to M$$

at $z_0 \in \mathbb{C}P^1$ is defined by $e_0(u, J) = u(z_0)$.

Theorem 6.1.1 *For every point $z_0 \in \mathbb{C}P^1$ the map $e_0 : \mathcal{M}(A, \mathcal{J}) \to M$ is a submersion.*

Recall from Section 3.4 that the tangent space of $\mathcal{M}(A, \mathcal{J})$ at a pair (u, J) with $\bar{\partial}_J(u) = 0$ is the set of all pairs

$$(\xi, Y) \in C^\infty(u^*TM) \times C^\infty(\mathrm{End}(TM, J, \omega))$$

which satisfy

$$D_u\xi + \tfrac{1}{2}Y(u)du \circ i = 0.$$

The differential of the evaluation map is obviously given by

$$de_0(u, J)(\xi, Y) = \xi(z_0).$$

We shall prove that, given any vector $v \in T_{u(z_0)}M$ there exists a pair $(\xi, Y) \in T_{(u,J)}\mathcal{M}(A, \mathcal{J})$ such that $\xi(z_0) = v$. Moreover, we shall prove that ξ can be chosen with support in an arbitrarily small neighbourhood of z_0. In fact we shall first construct a local solution ξ of $D_u\xi = 0$ near the point z_0 and then modify ξ by a cutoff function to make it vanish outside a small neighbourhood of z_0. We then have to compensate for the effect of the cutoff function by introducing an infinitesimal almost complex structure Y which is supported in a neighbourhood of the image under u of a small annulus centered at z_0.

Lemma 6.1.2 *Given $J \in \mathcal{J}(M, \omega)$ and a J-holomorphic curve $u : \mathbb{C}P^1 \to M$ there exists a constant $\delta > 0$ such that for every $v \in T_{u(z_0)}M$ and every pair $0 < \rho < r < \delta$ there exists a smooth vector field $\xi(z) \in T_{u(z)}M$ along u and an infinitesimal almost complex structure $Y \in C^\infty(\mathrm{End}(TM, J, \omega))$ such that the following holds.*

(i) $D_u\xi + \tfrac{1}{2}Y(u)du \circ i = 0.$

(ii) $\xi(z_0) = v.$

(iii) *ξ is supported in $B_r(z_0)$ and Y is supported in an arbitrarily small neighbourhood of $u(B_r(z_0) - B_\rho(z_0))$.*

Proof: Recall from the proof of Proposition 3.4.1 that

$$D_u\xi + \tfrac{1}{2}Y(u)du \circ i = \eta\, ds - J(u)\eta\, dt$$

where

$$\eta = \partial_s\xi + J(u)\partial_t\xi + (\partial_\xi J(u))\partial_t u + Y(u)\partial_t u.$$

Here we denote by $z = s + it$ conformal coordinates on $\mathbb{C}P^1$ and think of $\xi(s, t) \in \mathbb{R}^{2n}$ as the local coordinate representation of a tangent vector in TM with respect to a Darboux coordinate chart. Now choose a unitary trivialization $\Phi(s, t) : \mathbb{R}^{2n} \to T_{u(s,t)}M$ such that

$$J(u(s,t))\Phi(s,t) = \Phi(s,t)J_0, \quad \omega(\Phi v, \Phi w) = (J_0 v)^T w$$

where

$$J_0 = \begin{pmatrix} 0 & -\mathbb{1} \\ \mathbb{1} & 0 \end{pmatrix},$$

and T denotes the transpose, and write

$$\xi(s,t) = \Phi(s,t)\xi_0(s,t), \qquad v = \Phi(0,0)v_0, \qquad \partial_t u(s,t) = -\Phi(s,t)\zeta_0(s,t).$$

Then we have

$$D_u\xi + \tfrac{1}{2}Y(u)du \circ i = \Phi(\eta_0\, ds - J(u)\eta_0\, dt)$$

where
$$\eta_0 = \partial_s \xi_0 + J_0 \partial_t \xi_0 + A_0 \xi_0 - Y_0 \zeta_0$$
and
$$Y_0 = \Phi^{-1} Y(u) \Phi, \qquad A_0 = \Phi^{-1} \left(\partial_s \Phi + J(u) \partial_t \Phi + (\partial_\Phi J)(u) \right).$$

The compatibility condition $Y(x) \in \mathrm{End}(T_x M, J_x, \omega_x)$ for $x \in M$ translates into
$$Y_0 = Y_0^T = J_0 Y_0 J_0.$$

We must find a matrix-valued function $Y_0 : B_1 \to \mathbb{R}^{2n \times 2n}$ and a vector field $\xi_0 : B_1 \to \mathbb{R}^{2n}$ such that
$$\partial_s \xi_0 + J_0 \partial_t \xi_0 + A_0 \xi_0 = Y_0 \zeta_0, \qquad \xi_0(0) = v_0.$$

First note that the condition $\partial_s \xi_0 + J_0 \partial_t \xi_0 = 0$ means that ξ_0 is holomorphic and there exists a holomorphic curve through every point v_0 (for example the constant function). Now a standard perturbation argument in Fredholm theory shows that if A_0 is sufficiently small then the equation
$$\partial_s \xi_0 + J_0 \partial_t \xi_0 + A_0 \xi_0 = 0$$

with $\xi_0(0) = v_0$ has a solution $\xi_0 : B_1 \to \mathbb{R}^{2n}$ in the unit ball for every value of v_0. For example this can be proved by considering a regular boundary value problem. By a rescaling argument this can be extended to arbitrary A_0 but with the domain of the solution ξ_0 restricted to B_δ where δ depends on A_0. Now for $0 < \rho < r < \delta$ choose a cutoff function
$$\beta : \mathbb{C} \to [0, 1]$$

such that $\beta(z) = 1$ for $|z| < \rho$ and $\beta(z) = 0$ for $|z| > r$. Then the function $\beta \xi_0$ satisfies
$$\partial_s (\beta \xi_0) + J_0 \partial_t (\beta \xi_0) + A_0 (\beta \xi_0) = (\partial_s \beta) \xi_0 + (\partial_t \beta) J_0 \xi_0 \stackrel{\mathrm{def}}{=} \eta_0.$$

So we must find Y_0 such that
$$Y_0 \zeta_0 = \eta_0$$

where $\zeta_0 = -\Phi^{-1} \partial_t u$. Such Y_0 exists because each space $\mathrm{End}(\mathbb{R}^{2n}, J_0, \omega_0)$ acts transitively on the vectors in $\mathbb{R}^{2n}$. An explicit formula is given by
$$\begin{aligned}
Y_0 \;=\; & \frac{1}{|\zeta_0|^2} \left(\eta_0 \zeta_0^T + \zeta_0 \eta_0^T \right) + \frac{1}{|\zeta_0|^2} J_0 \left(\eta_0 \zeta_0^T + \zeta_0 \eta_0^T \right) J_0 \\
& - \frac{\langle \eta_0, \zeta_0 \rangle}{|\zeta_0|^4} \left(\zeta_0 \zeta_0^T + J_0 \zeta_0 \zeta_0^T J_0 \right) \\
& - \frac{\langle \eta_0, J_0 \zeta_0 \rangle}{|\zeta_0|^4} \left(J_0 \zeta_0 \zeta_0^T - \zeta_0 \zeta_0^T J_0 \right).
\end{aligned}$$

This function is well defined wherever $du(s,t) \neq 0$ and satisfies $Y_0^T = Y_0 = J_0 Y_0 J_0$ and $Y_0 \zeta_0 = \eta_0$. Now it follows from Lemma 2.2.3 that the restriction of u to an (arbitrarily small) annulus $B_r - B_\rho$ is an embedding.[1] For such a choice of ρ and r there exists a $Y \in C^\infty(\mathrm{End}(TM, J, \omega))$ such that
$$\Phi(s,t) Y_0(s,t) = Y(u(s,t)) \Phi(s,t).$$

---

[1] This is where we use the hypothesis that u is simple.

This function Y has the required properties and is supported in an arbitrarily small neighbourhood of $u(B_r - B_\delta)$. This proves the lemma. $\square$

Theorem 6.1.1 follows immediately from Lemma 6.1.2. An alternative geometric proof can be given in terms of the local deformation theory for J-holomorphic curves as in Nijenhuis and Woolf [57] or McDuff [44], [48], [50]: see [42, Proposition 4.1]. In fact our proof of Lemma 6.1.2 can be viewed as a linearized version of these deformation results.

6.2 Moduli spaces of N-tuples of curves

Consider the universal moduli space

$$\mathcal{M}(A^1, \dots, A^N, \mathcal{J})$$

of N-tuples of distinct curves. This consists of all $(N+1)$-tuples $(u^1, \dots, u^N, J)$ such that $J \in \mathcal{J}$, $u^j \in \mathcal{M}(A^j, J)$ and

$$u^i \neq u^j \circ \phi \quad \text{for} \ \phi \in \mathrm{G} \ \text{and} \ i \neq j.$$

In the case $A^i \neq A^j$ the condition $u^i \neq u^j \circ \phi$ is automatically satisfied and the set $\mathcal{M}(A^1, \dots, A^N, \mathcal{J})$ is simply the union of the sets of all N-tuples $(u^1, \dots, u^N)$ with $u^j \in \mathcal{M}(A^j, J)$ over all $J \in \mathcal{J}$. In the case $A^i = A^j$ for some pair $i \neq j$ we have deleted the "diagonal" of all N-tuples with $u^i = u^j \circ \phi$ from the moduli space. Note that if $u^i = u^j \circ \phi$ for some $i \neq j$ the corresponding cusp-curve $\{C^1, \dots, C^N\}$ (where $C^i = u^i(\mathbb{C}P^1)$) is not simple and hence will not appear in the proof of Theorem 5.2.1 or 5.3.1.

Remark 6.2.1 It follows from Proposition 2.3.2 that the reparametrization group $\mathrm{G} \times \cdots \times \mathrm{G} = \mathrm{G}^N$ acts freely on the space $\mathcal{M}(A^1, \dots, A^N, \mathcal{J})$ by

$$\phi \cdot (u^1, \dots, u^N, J) = (u^1 \circ \phi_1^{-1}, \dots, u^N \circ \phi_N^{-1}, J)$$

for $\phi = (\phi_1, \dots, \phi_N) \in \mathrm{G}^N$. $\square$

Following the line of argument in Chapter 3 we introduce the space $\mathcal{J}^\ell$ of almost complex structures of class C^ℓ where $\ell \geq 1$. We also choose $p > 2$ and $1 \leq k \leq \ell + 1$ and denote by

$$\mathcal{M}^\ell(A^1, \dots, A^N, \mathcal{J}) = \mathcal{M}(A^1, \dots, A^N, \mathcal{J}^\ell)$$

the corresponding universal moduli space where $J \in \mathcal{J}^\ell$ and the J-holomorphic curves u^j are of class $W^{k,p}$. Recall from Chapter 3 that this space is independent of the choice of k and p.

Proposition 6.2.2 *The space $\mathcal{M}^\ell(A^1, \dots, A^N, \mathcal{J})$ is a Banach manifold.*

Proof: The proof is almost word by word the same as that of Proposition 3.4.1 and we shall content ourselves with summarizing the main points. Denote by $\mathcal{X}$ the space of all smooth maps

$$u = (u^1, \dots, u^N) : \mathbb{C}P^1 \to M^N$$

for which there exists an N-tuple $(z_1, \ldots, z_N) \in (\mathbb{C}P^1)^N$ which satisfies the conditions of Proposition 2.3.2. This set is open (with respect to the C^1-topology) in the space of all smooth maps from $\mathbb{C}P^1$ to M^N and it follows from Proposition 2.3.2 that $\mathcal{M}(A^1, \ldots, A^N, \mathcal{J}) \subset \mathcal{X}$. In fact, our universal moduli space is the set of zeros of a section of the bundle $\mathcal{E} \to \mathcal{X}$ with fibers

$$\mathcal{E}_u = \Omega^{0,1}((u^1)^*TM) \oplus \cdots \oplus \Omega^{0,1}((u^N)^*TM)$$

This section $\mathcal{F} : \mathcal{X} \to \mathcal{E}$ is given by

$$\mathcal{F}(u^1, \ldots, u^N, J) = (\bar{\partial}_J(u^1), \ldots, \bar{\partial}_j(u^N)).$$

Now the differential of this section at a zero (u, J) is the operator

$$D\mathcal{F}(u, J) : T_u\mathcal{X} \oplus T_J\mathcal{J} \to \mathcal{E}_u$$

given by

$$D\mathcal{F}(u, J)(\xi, Y) = (D_{u^1}\xi_1 + \tfrac{1}{2}Y \circ du^1 \circ i, \ldots, D_{u^N}\xi_N + \tfrac{1}{2}Y \circ du^N \circ i).$$

where $\xi = (\xi_1, \ldots, \xi_N) \in T_u\mathcal{X} = C^\infty((u^1)^*TM) \oplus \cdots \oplus C^\infty((u^N)^*TM)$. The proof that this operator is onto (when considered on the usual Sobolev spaces) is word by word the same as that of Proposition 3.4.1. The only point of difference is that the existence of an N-tuple $(z_1, \ldots, z_N) \in (\mathbb{C}P^1)^N$ which satisfies the conditions of Proposition 2.3.2 is used to prove that the operator $D\mathcal{F}(u, J)$ has a dense range. Further details are left to the reader. $\qquad\qquad\square$

Observe that if $A^i = A^j$ for some $i \neq j$ then we can add a point $(u, J) = (u^1, \ldots, u^N, J)$ with $u^i = u^j \circ \phi$ to the moduli space without destroying the manifold structure provided that the operator $D\mathcal{F}(u, J)$ is onto. For example, this will be the case if the operators D_{u^j} are onto for all j. However, adding such curves would destroy the free action of the group G^N on $\mathcal{M}(A^1, \ldots, A^N, \mathcal{J})$.

6.3   Moduli spaces of cusp-curves

Recall that a cusp-curve C which represents the class A is a connected union

$$C = C^1 \cup \ldots \cup C^N$$

of J-holomorphic curves $C^j = u^j(\mathbb{C}P^1)$, which are called its components, such that the corresponding homology classes $A^j = [C^j]$ have the sum $A^1 + \cdots + A^N = A$. Since C is reducible, either $N > 1$ or C^1 is multiply covered. C is said to be simple if all its components are distinct and none are multiply-covered. Any curve C can be simplified by deleting all but one copy of repeated components, and by replacing any multiply-covered component by its underlying simple curve. Note that both these operations change the homology class $[C]$. We will see that to compactify the set $\mathcal{C}(A, J)$ of all simple A-curves we will need to add to it all simplified cusp-curves which represent A. The following lemma was already used in Remark 4.4.2. The proof is an elementary exercise.

Lemma 6.3.1 *The components of C can be ordered so that $C^1 \cup \ldots \cup C^k$ is connected for all $k \leq N$.*

Each simple cusp-curve $C = C^1 \cup \ldots \cup C^N$ has a type D. This is the set $\{A^1, \ldots, A^N\}$ of homology classes $A^j = [C^j]$, together with a framing which specifies that certain components of C intersect. We do not need to describe the full intersection pattern as was done in [64], but just enough of it to ensure that C is connected. Now suppose that the components of of C are ordered as in Lemma 6.3.1. Then each component C^ν must intersect some component C^{j_ν} where $1 \leq j_\nu \leq \nu - 1$, and so we take the set of integers $\{j_2, \ldots, j_N\}$ to be the framing data. For every framed class

$$D = \{A^1, \ldots, A^N, j_2, \ldots, j_N\}$$

and each $J \in \mathcal{J}(M, \omega)$ denote by

$$\mathcal{M}(D, J) \subset \mathcal{M}(A^1, \ldots, A^N, J) \times (\mathbb{C}P^1)^{2N-2}$$

the moduli space of all points (u, w, z) with $w = (w_2, \ldots, w_N) \in (\mathbb{C}P^1)^{N-1}$, $z = (z_2, \ldots, z_N) \in (\mathbb{C}P^1)^{N-1}$, and

$$u^j \in \mathcal{M}(A^j, J)$$

such that $u = (u^1, \ldots, u^N)$ is a simple J-holomorphic cusp-curve with

$$u^{j_\nu}(w_\nu) = u^\nu(z_\nu)$$

for $\nu = 2, \ldots, N$. Our first goal in this section is to prove that for a generic choice of the almost complex structure the moduli space $\mathcal{M}(D, J)$ is a smooth finite dimensional manifold.

Theorem 6.3.2 *There exists a set $\mathcal{J}_{\mathrm{reg}} = \mathcal{J}_{\mathrm{reg}}(D) \subset \mathcal{J}$ of the second category such that $\mathcal{M}(D, J)$ is a manifold of dimension*

$$\dim \mathcal{M}(D, J) = 2 \sum_{j=1}^{N} c_1(A^j) + 2n + 4(N - 1).$$

for every $J \in \mathcal{J}_{\mathrm{reg}}$.

As before the letter $\mathcal{J}$ denotes the space of ω-compatible almost complex structures.

To prove this theorem consider the evaluation map

$$e_D : \mathcal{M}(A^1, \ldots, A^N, J) \times (\mathbb{C}P^1)^{2N-2} \to M^{2N-2}$$

defined by

$$e_D(u, w, z) = (u^{j_2}(w_2), u^2(z_2), \ldots, u^{j_N}(w_N), u^N(z_N))$$

for $u \in \mathcal{M}(A^1, \ldots, A^N, \mathcal{J})$, $w = (w_2, \ldots, w_N) \in (\mathbb{C}P^1)^{N-1}$, and $z = (z_2, \ldots, z_N) \in (\mathbb{C}P^1)^{N-1}$. With this notation the moduli space $\mathcal{M}(D, J)$ is the inverse image of the multi-diagonal

$$\Delta_N = \left\{ (x, y) = (x_2, y_2, \ldots, x_N, y_N) \in M^{2N-2} \,\middle|\, x_\nu = y_\nu \right\}$$

under the evaluation map e_D. Thus, Theorem 6.3.2 asserts that the set $\mathcal{J}_{\text{reg}}(D)$ of J for which the evaluation map e_D is transverse to Δ_N has second category in $\mathcal{J}(M, \omega)$. To prove this we consider the *universal moduli space*

$$\mathcal{M}(D, \mathcal{J})$$

of all points (u, J, w, z) with $J \in \mathcal{J}(M, \omega)$ and $(u, w, z) \in \mathcal{M}(D, J)$. This space is defined to be the inverse image of the diagonal Δ_N under the extended evaluation map

$$e_D : \mathcal{M}(A^1, \ldots, A^N, \mathcal{J}) \times (\mathbb{C}P^1)^{2N-2} \to M^{2N-2}$$

which is defined by the same formula as above. For the case $N = 2$ the following crucial transversality result was proved by McDuff in [42].

Proposition 6.3.3 *For each framed class* $D = \{A^1, \ldots, A^N, j_2, \ldots, j_N\}$ *the evaluation map*

$$e_D : \mathcal{M}(A^1, \ldots, A^N, \mathcal{J}) \times (\mathbb{C}P^1)^{2N-2} \to M^{2N-2}$$

is transversal to the multi-diagonal $\Delta_N \subset M^{2N-2}$. *Hence*

$$\mathcal{M}(D, \mathcal{J}) = e_D^{-1}(\Delta_N).$$

is an infinite dimensional Fréchet manifold.

Proof: Fix a $(2N - 2)$-tuple $(w, z) \in (\mathbb{C}P^1)^{2N-2}$. It suffices to prove that the map

$$\mathcal{M}(A^1, \ldots, A^N, \mathcal{J}) \to M^{2N-2} : (u, J) \mapsto e_D(u, J, w, z)$$

is transverse to Δ_N. The tangent space to $\mathcal{M}(A^1, \ldots, A^N, \mathcal{J})$ at the point $(u, J) = (u^1, \ldots, u^N, J)$ consists of all $(\xi, Y) = (\xi_1, \ldots, \xi_N, Y)$ where

$$\xi_\nu \in C^\infty((u^\nu)^* TM), \qquad Y \in C^\infty(\text{End}(TM, J, \omega)),$$

and

$$D_{u^\nu} \xi_\nu + \tfrac{1}{2} Y(u^\nu) du^\nu \circ i = 0.$$

The differential of the map $(u, J) \mapsto e_D(u, J, w, z)$ at a point $(u, J, w, z) \in e_D^{-1}(\Delta_N)$ is the map

$$T_{(u,J)} \mathcal{M}(A^1, \ldots, A^N, \mathcal{J}) \to \bigoplus_{\nu=2}^N T_{x_\nu} M \oplus T_{x_\nu} M$$

which assigns to the $(N + 1)$-tuple $(\xi_1, \ldots, \xi_N, Y)$ the $(2N - 2)$-tuple $(\widehat{x}, \widehat{y}) = (\widehat{x}_2, \ldots, \widehat{x}_N, \widehat{y}_2, \ldots, \widehat{y}_N)$ given by

$$\widehat{x}_\nu = \xi_{j_\nu}(w_\nu), \qquad \widehat{y}_\nu = \xi_\nu(z_\nu).$$

Here we denote $x_\nu = u^{j_\nu}(w_\nu) = u^\nu(z_\nu)$ and so $\widehat{x}_\nu, \widehat{y}_\nu \in T_{x_\nu} M$. Given $\widehat{y}_\nu$ it follows from Lemma 6.1.2 that there exists a $Y \in C^\infty(\text{End}(TM, J, \omega))$ and a $\xi_\nu \in C^\infty(u_\nu^* TM)$ such that

$$\xi_\nu(z_\nu) = \widehat{y}_\nu.$$

Moreover, ξ_ν can be chosen with support in a small small neighbourhood B_ν of z_ν and Y can be chosen with support in an arbitrarily small neighbourhood of

$u^\nu(B'_\nu - B_\nu)$ where $B'_\nu - B_\nu$ is a small annulus centered at z_ν By Proposition 2.3.2 these annuli can be chosen such that $u^\nu(B'_\nu - B_\nu)$ does not intersect any of the other J-holomorphic curves $w^j(\mathbb{C}P^1)$ for $j \neq \nu$. Moreover, if $w_\nu \neq z_{j_\nu}$ then the ball B'_{j_ν} can be chosen so small that $w_\nu \notin B'_{j_\nu}$ and hence

$$w_\nu \neq z_{j_\nu} \qquad \Longrightarrow \qquad \widehat{x}_\nu = \xi_{j_\nu}(w_\nu) = 0.$$

If on the other hand $w_\nu = z_{j_\nu}$ then we have

$$w_\nu = z_{j_\nu} \qquad \Longrightarrow \qquad \widehat{x}_\nu = \xi_{j_\nu}(w_\nu) = \widehat{y}_{j_\nu}.$$

Now it is a simple combinatorial exercise that the space of all $(2N - 2)$-tuples $(\widehat{x}_2, \dots, \widehat{x}_N, \widehat{y}_2, \dots, \widehat{y}_N)$ with arbitrary $\widehat{y}_\nu$ and $\widehat{x}_\nu$ given by the above conditions is transverse to the diagonal. This proves the proposition. $\square$

Proof of Theorem 6.3.2:   The proof is essentially an application of the Sard-Smale theorem. As always, this theorem applies only to maps between Banach manifolds and so one must work with the space $\mathcal{J}^\ell$ of almost complex structures of class C^ℓ and a suitable Sobolev space of J-holomorphic curves, as in the proof of Theorem 3.1.2. These details will be left to the reader from now on.

Consider the projection

$$\pi_D : \mathcal{M}(D, \mathcal{J}) \to \mathcal{J}.$$

We first show that this map is Fredholm. To see this note that π_D is a composite

$$\mathcal{M}(D, \mathcal{J}) \to \mathcal{M}(A^1, \mathcal{J}) \to \mathcal{J},$$

where the second map π_{A^1} is Fredholm. Therefore, it suffices to consider the first map. But this has derivative

$$(\xi, Y, \widehat{w}, \widehat{z}) \mapsto (\xi_1, Y),$$

where $(\xi_j, Y) \in T_{(u^j, J)}\mathcal{M}(A^j, \mathcal{J})$ for $j = 1, \dots, N$. Hence its kernel and cokernel are finite dimensional. Therefore π_D is Fredholm, and so its fibers $\mathcal{M}(D, J)$ are manifolds for generic J.

We shall now prove that (u, J, w, z) is a regular point for π_D if and only if at this point the restricted evaluation map

$$e_D : \mathcal{M}(A^1, \dots, A^N, J) \times (\mathbb{C}P^1)^{2N-2} \to M^{2N-2}$$

is transverse to the diagonal. But that (u, J, w, z) is a regular point for $\pi_D :$ $\mathcal{M}(D, \mathcal{J}) \to \mathcal{J}$ just means that for every $Y \in T_J\mathcal{J}$ there exists $(\xi, \widehat{w}, \widehat{z})$ such that $(\xi, Y, \widehat{w}, \widehat{z}) \in T_{(u, J, w, z)}\mathcal{M}(D, \mathcal{J})$ or equivalently

$$de_D(u, J, w, z)(\xi, Y, \widehat{w}, \widehat{z}) \in T_{e_D(u, J, w, z)}\Delta_N.$$

Since, by Proposition 6.3.3, the vectors

$$de_D(u, J, w, z)(\xi, Y, \widehat{w}, \widehat{z})$$

with $\xi_j \in T_{u^j}\mathcal{M}(A^j, J)$ are transverse to the multi-diagonal Δ_N, it is easy to check that the subspace of all such vectors with $Y = 0$ must already be transverse to Δ_N. Thus we have proved that J is a regular value for the projection $\pi_D : \mathcal{M}(D, \mathcal{J}) \to \mathcal{J}$ if and only if $e_D : \mathcal{M}(A^1, \dots, A^N, J) \times (\mathbb{C}P^1)^{2N-2} \to M^{2N-2}$ is transverse to the diagonal. This proves the theorem. $\square$

6.4 Evaluation maps for cusp-curves

Note that the dimension of the moduli space $\mathcal{M}(D, J)$ with $A^1 + \cdots + A^N = A$ is $2c_1(A) + 2n + 4(N - 1)$ and hence is equal to that of $\mathcal{M}(A, J) \times (\mathbb{C}P^1)^{2N-2}$. However, this is deceptive, since the $6N$-dimensional group $\mathrm{G}^N = \mathrm{G} \times \cdots \times \mathrm{G}$ acts on this space by

$$\phi \cdot (u^j, w_\nu, z_\nu) = (u^j \circ \phi_j^{-1}, \phi_{j_\nu}(w_\nu), \phi_\nu(z_\nu)).$$

The space of simple unparametrized curves of type D is denoted by

$$\mathcal{C}(D, J) = \mathcal{M}(D, J)/\mathrm{G}^N.$$

By Proposition 2.3.2 the group G^N acts freely on this space. Hence the quotient is again a manifold for $J \in \mathcal{J}_{\mathrm{reg}}(D)$ and its dimension is

$$\dim \mathcal{C}(D, J) = 2c_1(A^1 + \cdots + A^N) + 2n - 2N - 4.$$

Note that in the case $N = 1$ and $A^1 = A$ this is precisely the dimension of the space $\mathcal{C}(A, J) = \mathcal{M}(A, J)/\mathrm{G}$ of unparametrized simple A-curves, namely $2c_1(A) + 2n - 6$. In general, however the dimension of the space $\mathcal{C}(D, J)$ will be at least two less than that of $\mathcal{C}(A, J)$ and this observation lies at the heart of Theorem 5.2.1.

In order to define the domain $\mathcal{W}(D, J)$ of the corresponding evaluation map we must quotient $\mathcal{M}(D, J)$ out by as large a reparametrization group as we can in order to reduce the dimension as much as possible. However, one component must be kept parametrized so that it can be used to define an evaluation map. Therefore fix $\ell \leq N$ and let $\mathrm{G}_{\ell,N}$ be the subgroup

$$\mathrm{G}_{\ell,N} = \{\phi = (\phi_1, \ldots, \phi_N) \mid \phi_\ell = \mathrm{id}\} \subset \mathrm{G}^N.$$

Then the space

$$\mathcal{C}_\ell(D, J) = \mathcal{M}(D, J)/\mathrm{G}_{\ell,N}.$$

consists of all simplified cusp-curves C of type D whose components, except for the ℓ-th, are unparametrized. It follows that G acts on $\mathcal{C}_\ell(D, J)$ by acting on its ℓ-th component in the usual way and so we define

$$\mathcal{W}_\ell(D, J) = \mathcal{C}_\ell(D, J) \times_{\mathrm{G}} \mathbb{C}P^1$$

as the quotient space under the obvious action and

$$\mathcal{W}(D, J) = \bigcup_\ell \mathcal{W}_\ell(D, J).$$

Again it follows from Proposition 2.3.2 that G acts freely on $\mathcal{C}_\ell(D, J) \times \mathbb{C}P^1$ and hence the quotient space $\mathcal{W}_\ell(D, J)$ is a manifold. There is, of course, an evaluation map $e : \mathcal{W}(D, J) \to M$ which is defined on $\mathcal{W}_\ell(D, J)$ by

$$e(u^j, w_\nu, z_\nu, \zeta) = u^\ell(\zeta).$$

Lemma 6.4.1 *Assume that $A \in H_2(M, \mathbb{Z})$ is not a multiple class mB where $m > 1$ and $c_1(B) = 0$. Let D be the type of a simplified cusp-curve which represents the class A. Suppose that $J \in \mathcal{J}^+(M, \omega, K) \cap \mathcal{J}_{\mathrm{reg}}(D)$ is K-semi-positive for some $K > \omega(A)$. Then $\mathcal{W}(D, J)$ is a manifold of dimension*

$$\dim \mathcal{W}(D, J) \leq \dim \mathcal{W}(A, J) - 2 \max\{1, (N-1)\}$$

for every $D \neq A$.

Proof: It just remains to check the statement about the dimension. First we have

$$
\begin{aligned}
\dim \mathcal{C}_\ell(D, J) &= \dim \mathcal{M}(A^1, \ldots, A^N, J) + \dim (\mathbb{C}P^1)^{2N-2} \\
&\quad - \dim \mathrm{G}_{\ell,N} - \mathrm{codim}\, \Delta_N \\
&= \sum_j 2c_1(A^j) + 2nN \\
&\quad + 4(N-1) - 6(N-1) - 2n(N-1) \\
&= 2n + \sum_j 2c_1(A^j) - 2(N-1).
\end{aligned}
$$

Now our hypotheses imply that either $N > 1$ or $D = \{A^1\}$ where $A = mA^1$ and $c_1(A^1) > 0$. In the first case, because $c_1(A^j) \geq 0$,

$$\sum_j c_1(A^j) \leq c_1(A).$$

Thus, in either case,

$$
\begin{aligned}
\dim \mathcal{W}(D, J) &= 2n + \sum_j 2c_1(A^j) - 2(N+1) \\
&\leq 2n + 2c_1(A) - 4 - 2\max\{1, (N-1)\} \\
&= \dim \mathcal{W}(A, J) - 2\max\{1, (N-1)\}.
\end{aligned}
$$

This proves the lemma. $\square$

Example 6.4.2 In the case of a cusp-curve with two components the description of the compactified moduli space can be slightly simplified. In this case we may fix a point $z_0 \in \mathbb{C}P^1$ and define

$$
\begin{aligned}
\mathcal{M}_0(A, B, J) = \{(u, v)\, |\, u \in \mathcal{M}(A, J),\ v \in \mathcal{M}(B, J),\ u(z_0) = v(z_0), \\
u \neq v \circ \phi\}
\end{aligned}
$$

For generic J this space is a manifold of dimension $2c_1(A + B) + 2n$. The 8-dimensional group $\mathrm{G}_0 \times \mathrm{G}_0$ acts on $\mathcal{M}_0(A, B, J)$ where $\mathrm{G}_0 = \{\phi \in G\, |\, \phi(z_0) = z_0\}$. Thus the space

$$\mathcal{C}(A, B, J) = \frac{\mathcal{M}_0(A, B, J)}{\mathrm{G}_0 \times \mathrm{G}_0}$$

of unparametrized cusp-curves of type (A, B) has dimension $2c_1(A + B) + 2n - 8$, which is 2 less than the dimension of the space $\mathcal{C}(A + B, J) = \mathcal{M}(A + B, J)/\mathrm{G}$ of unparametrized curves of type $A + B$. Now the natural domain of the evaluation map can be defined as

$$\mathcal{W}_0(A, B, J) = \frac{\mathcal{M}_0(A, B, J)}{G_{01} \times G_0} \cup \frac{\mathcal{M}_0(A, B, J)}{G_0 \times G_{01}}$$

where $G_{01} = \{\phi \in G_0 \mid \phi(z_1) = z_1\}$ for some $z_1 \neq z_0$. Both groups $G_{01} \times G_0$ and $G_0 \times G_{01}$ act freely on $\mathcal{M}_0(A, B, J)$ and hence

$$\dim \mathcal{W}_0(A, B, J) = 2c_1(A + B) + 2n - 6.$$

Again this is 2 less than the dimension of $\mathcal{M}(A+B, J) \times_G CP^1$. The evaluation map $\mathcal{W} \to M$ is given by $(u, v) \mapsto u(z_1)$ on the first component, and by $(u, v) \mapsto v(z_1)$ on the second. It is easy to check that the space $\mathcal{W}_0(A, B, J)$ is diffeomorphic to the above space $\mathcal{W}(D, J)$ for $D = \{A, B\}$. $\qquad\square$

6.5 Proofs of the theorems in Sections 5.2 and 5.3

We are now in a position to prove Theorems 5.2.1 and 5.3.1 on compactifying the image of evaluation maps.

Proof of Theorem 5.2.1: By Gromov's compactness theorem 4.4.3, the closure of the set $X(A, J)$ will contain points in M which lie on some cusp-curve which represents the class A. Clearly, any such curve may be simplified, without changing the set of points which lie on it, by deleting any repeated components and replacing multiply covered components by their underlying simple components. Because cusp-curves are connected the resulting simplified cusp-curve will lie in some set $C(D, J)$.

Thus, in order to compactify $X(A, J)$ it suffices to add the points in $X(D, J)$, as D ranges over all effective, framed classes which can represent a simplified J-holomorphic cusp-curves whose energy is bounded by $\omega(A)$. By Corollary 4.3.3, there are only finitely many of these. This corollary also implies that the set of D which must be considered for each J is locally constant as J varies in $\mathcal{J}$. This proves (i). Statements (ii) and (iii) follow from Lemma 6.4.1. The important observation here is that $c_1(A^j) \geq 0$ for all j and therefore simplification of a cusp-curve decreases the Chern number and hence the dimension of the space $\mathcal{W}(D, J)$. $\square$

Proof of Theorem 5.3.1: This is essentially the same as the proof of Theorem 5.2.1. The only challenge is to define the manifolds $\mathcal{W}(D, T, J, p)$ correctly. Recall that the map

$$T : \{1, \ldots, p\} \to \{1, \ldots, N\}$$

describes which of the N components of C are evaluated to obtain a p-tuple in M^p. Hence we define

$$X(D, T, J, p) = \left\{(x_1, \ldots, x_p) \mid x_j \in C_{T(j)}, C \in C(D, J)\right\},$$

and set

$$\mathcal{W}(D, T, J, p) = \mathcal{M}(D, J) \times_{G^N} (\mathbb{C}P^1)^p,$$

where the ν-th factor ϕ_ν of of $\phi = (\phi_1, \ldots, \phi_N) \in G^N$ acts on $\mathcal{M}(D, J)$ as before, and acts on the j-th factor of $\mathbb{C}P^1$ if and only if $T(j) = \nu$. Then there is an evaluation map

$$e_{D,T} : \mathcal{W}(D, T, J, p) \to M^p$$

defined by

$$e_{D,T}(u, J, w, z, \zeta) = (u_{T(1)}(\zeta_1), \ldots, u_{T(p)}(\zeta_p))$$

for $(u, J, w, p) \in \mathcal{M}(D, J)$ and $\zeta = (\zeta_1, \ldots, \zeta_p) \in (\mathbb{C}P^1)^p$. Now the dimension of $\mathcal{W}(D, T, J, p)$ is given by

$$\begin{aligned}
\dim \mathcal{W}(D, T, J, p) &= \dim \mathcal{M}(D, J) + 2p - 6N \\
&= 2 \sum_{j=1}^{N} c_1(A^j) + 2n + 4(N - 1) + 2p - 6N \\
&= 2 \sum_{j=1}^{N} c_1(A^j) + 2n + 2p - 2N - 4.
\end{aligned}$$

In the case $N = 1$ with $D = \{A\}$ this number agrees with the dimension of the domain of the p-fold evaluation map on the space of simple curves

$$\dim \mathcal{M}(A, J) \times_G (\mathbb{C}P^1)^p = 2c_1(A) + 2n + 2p - 6.$$

But in the case $N \geq 2$ or $N = 1$ with $A = mA^1$ for $m \geq 2$ the the dimension of $\mathcal{W}(D, T, J, p)$ is at least two less than $2c_1(A) + 2n + 2p - 6$. In view of this all the statements in Theorem 5.3.1 follow from the corresponding statements in the case $p = 1$. $\qquad\square$

The following lemma gives some useful information about the top dimensional pieces in the *boundary* $\partial \mathcal{C}(A, J) = \overline{\mathcal{C}}(A, J) - \mathcal{C}(A, J)$ of the moduli space of simple curves.

Lemma 6.5.1 *Assume that $A \in H_2(M, \mathbb{Z})$ is not a multiple class mB where $m > 1$ and $c_1(B) = 0$ and suppose that $J \in \mathcal{J}^+(M, \omega, K) \cap \mathcal{J}_{\mathrm{reg}}(D)$ is K-semi-positive for some $K > \omega(A)$. Then the codimension 2 pieces in $\partial \mathcal{C}(A, J)$ consist of curves of the following types:*

(i) *simple cusp-curves of type $D = \{A^1, A^2\}$ where $c_1(A^1 + A^2) = c_1(A)$;*

(ii) *simple B-curves where $c_1(A) = 2$ and $A = 2B$.*

Proof: This follows from Lemma 6.4.1. Note that in (i) we must have $m_1 A^1 + m_2 A^2 = A$ for some $m_1, m_2 > 0$, and hence we will have $m_i = 1$ unless $c_1(C^i) = 0$. $\square$

6.6 Proof of the theorem in Section 5.4

The definition of the spaces $\mathcal{V}(D, T, J, \mathbf{z})$ which stratify the boundary of the set $Y(A, J, p)$ is slightly more complicated than that of the spaces $\mathcal{W}(D, T, J, p)$, because we must take into account the interaction of the points z_i with the limiting process.

There are four cases for the possible limit behaviour of the points $e_{\mathbf{z}}(u_\nu)$ for a sequence $u_\nu \in \mathcal{M}(A, J)$. These are that u_ν converges modulo bubbling

(a) to a simple J-holomorphic sphere,

(b) to an m-fold covering of a simple J-holomorphic sphere where $2m < p$,

(c) to an m-fold covering of a simple J-holomorphic sphere where $2m \geq p$,

(d) to a constant map.

We shall treat the cases (a) and (b) simultaneously. In fact, (a) is just the special case $m = 1$. Let $X \subset \mathbb{C}P^1$ be the finite set at which holomorphic spheres bubble off. Then u_ν converges on the complement of X to a curve $v^0 \circ \psi$ where

$$v^0 \in \mathcal{M}(A^0, J)$$

is a simple J-holomorphic curve representing the class A^0 and $\psi : \mathbb{C}P^1 \to \mathbb{C}P^1$ is a rational map of degree m. Denote the space of such maps by Rat_m. It has real dimension

$$\dim \mathrm{Rat}_m = 4m + 2.$$

Suppose that ℓ of the points z_j lie in the set X and assume without loss of generality that these are the points $z_1, \ldots, z_\ell$. Then we must consider limit curves of the form $(\psi, v^0, \ldots, v^N)$ with

$$v^\nu(0) = v^0(\psi(z_\nu)), \qquad \nu = 1, \ldots, \ell.$$

These equations express the fact that the first ℓ bubbles are attached at the images under $v^0 \circ \psi$ of the points $z_1, \ldots, z_\ell$. The remaining intersection pattern of the limiting curve is unconstrained, and can therefore be described by integers $j_{\ell+1}, \ldots, j_N$ with $0 \leq j_\nu \leq \nu - 1$ such that

$$v^\nu(0) = v^{j_\nu}(w_\nu), \qquad \nu = \ell + 1, \ldots, N,$$

for some points $w_\nu \in \mathbb{C}P^1$. Thus the intersection pattern of this curve can be coded in a framing D of the form

$$D = \{m, A^0, \ldots, A^n, j_{\ell+1}, \ldots, j_N\},$$

where there are positive integers $m_0, m_1, \ldots, m_N$ such that $m_0 = m$ and

$$\sum_{j=0}^{N} m_j A^j = A.$$

There is a corresponding moduli space $\mathcal{M}(D, J, \mathbf{z})$ which consists of all $2N+2$-tuples

$$(\psi, v^0, v^1, \ldots, v^N, w_{\ell+1}, \ldots, w_N)$$

satisfying these conditions, where v^j represents the class A^j.

Lemma 6.6.1 *For generic J, the moduli space $\mathcal{M}(D, J, \mathbf{z})$ is a manifold of dimension*

$$\dim \mathcal{M}(D, J, \mathbf{z}) = 2n + 2c_1(D) + 2N - 2\ell + 4m + 2.$$

Proof: Consider the evaluation map

$$e_D : \mathrm{Rat}_m \times \mathcal{M}(A^0, \ldots, A^N, \mathcal{J}) \times (\mathbb{C}P^1)^{N-\ell} \to M^N \times M^N$$

defined by

$$e_D(\psi, v^0, \ldots, v^N, J, w_{\ell+1}, \ldots, w_N) = (x, y)$$

with

$$x = (v^1(0), \ldots, v^N(0))$$

and

$$y = (v^0(\psi(z_1)), \ldots, v^0(\psi(z_\ell)), v^{j_{\ell+1}}(w_{\ell+1}), \ldots, v^{j_N}(w_N)).$$

It can be proved as in Proposition 6.3.3 that this map is transverse to the diagonal and hence the space $\mathcal{M}(D, \mathcal{J}, \mathbf{z}) = e_D^{-1}(\Delta)$ is a Fréchet manifold.

Now consider the projection

$$\mathcal{M}(D, \mathcal{J}, \mathbf{z}) \to \mathcal{J}.$$

As before the regular values of this projection are the points for which the restricted evaluation map

$$e_{D,J} : \mathrm{Rat}_m \times \mathcal{M}(A^0, \ldots, A^N, J) \times (\mathbb{C}P^1)^{N-\ell} \to M^N \times M^N$$

is transverse to the diagonal. For such values of J the space $\mathcal{M}(D, J, \mathbf{z}) = e_{D,J}^{-1}(\Delta)$ is a manifold of dimension

$$\begin{aligned} \dim \mathcal{M}(D, J, \mathbf{z}) &= \dim \mathcal{M}(A^0, \ldots, A^N, J) + 2(N - \ell) + 4m + 2 - 2nN \\ &= 2n(N + 1) + 2c_1(D) + 2(N - \ell) + 4m + 2 - 2nN \\ &= 2n + 2c_1(D) + 2N - 2\ell + 4m + 2. \end{aligned}$$

$$\square$$

Now choose a map $T : \{1, \ldots, \ell\} \to \{1, \ldots, N\}$ and define the space

$$\mathcal{V}(D, T, J, \mathbf{z}) = \frac{\mathcal{M}(D, J, \mathbf{z}) \times (\mathbb{C}P^1)^\ell}{\mathrm{G} \times \mathrm{G}_0^N}$$

where $\mathrm{G}_0 = \{\phi \in \mathrm{G} \,|\, \phi(0) = 0\}$, and the group $\mathrm{G} \times \mathrm{G}_0^N$ acts by

$$\phi \cdot (\psi, v, w, \zeta) = (\phi_0 \circ \psi, v^\nu \circ \phi_\nu^{-1}, \phi_{j_\nu}(w_\nu), \phi_{T(\nu)}(\zeta_\nu))$$

for $(\psi, v, w) \in \mathcal{M}(D, J, \mathbf{z})$ and $\zeta \in (\mathbb{C}P^1)^\ell$. The evaluation map is defined by

$$e_{D,T,\mathbf{z}}(\psi, v, w, \zeta) = (v^{T(1)}(\zeta_1), \ldots, v^{T(\ell)}(\zeta_\ell), v^0(\psi(z_{\ell+1})), \ldots, v^0(\psi(z_N))).$$

Note that the variables ζ_i parametrize the ith bubble, for $i \leq \ell$. These bubbles have to be added to the boundary of $Y(A, J, p)$ because the set of points $u_\nu(z_i)$ might accumulate on any point on this bubble. Observe also that the image of the points z_j for $j > \ell$ is not determined completely by the map v^0, but may be moved around by difference choices of the multiple covering $\psi : \mathbb{C}P^1 \to \mathbb{C}P^1$.

Lemma 6.6.2 *Under the hypotheses of Theorem 5.4.1(iii), the above manifolds* $\mathcal{V}(D, T, J, \mathbf{z})$ *have dimension strictly less than that of* $\mathcal{M}(A, J)$.

Proof: The group G_0 is 4-dimensional and hence the space $\mathcal{V}(D, T, J, \mathbf{z})$ has dimension

$$
\begin{aligned}
\dim \mathcal{V}(D, T, J, \mathbf{z}) &= \dim \mathcal{M}(D, \mathbf{z}, J) + 2\ell - 4N \\
&= 2n + 2c_1(D) - 2N + 4m - 4.
\end{aligned}
$$

(We have added 2ℓ to the dimension of $\mathcal{M}(D, J, \mathbf{z})$ to account for $(\mathbb{C}P^1)^\ell$ and subtracted $4N + 6$ for the dimension of $G \times G_0^N$.) If $m = 1$ then we must have $N \geq 1$ and in this case the dimension is at most $2n + 2c_1(A) - 2$ as claimed. Here we only need the condition $c_1(D) \leq c_1(A)$ which is guaranteed by semi-positivity. In fact,

$$
c_1(D) = \sum_{j=0}^{N} c_1(A^j) \leq c_1(A) - (m-1)c_1(A^0).
$$

Now suppose $m > 1$. Then by assumption of case (b) we have $p > 2m$ and so condition (JA_p) implies that $c_1(A^0) > 2$. Hence

$$
c_1(D) + 2m - 2 < c_1(A).
$$

This implies again that

$$
\dim \mathcal{V}(D, T, J, \mathbf{z}) < 2n + 2c_1(A) = \dim \mathcal{M}(A, J).
$$

Now we cannot assume in general that precisely the first ℓ points appear on the bubbles. So for the general case we have to allow for a permutation of the points $z_1, \ldots, z_p$ and compensate with the inverse permutation of the factors in M^p. Obviously such a permutation will not change the dimension count. $\qquad\square$

This takes care of limit points of the form (a) or (b). Now consider the case (c). This means that u_ν converges to a multiply covered J-holomorphic sphere of class mA^0 where

$$
p \leq 2m.
$$

The moduli space of corresponding simplified cusp-curves is determined by a framed class

$$
D = \{A^0, \ldots, A^N, j_1, \ldots, j_N\}
$$

with $(N + 1)$ components, and Chern number

$$
c_1(D) \leq c_1(A) - (m-1)c_1(A^0).
$$

Since $p \leq 2m$, the reparametrization map ψ acts transitively on p-tuples $\mathbf{z}$ of distinct points, and so the set of points $u_\nu(\mathbf{z})$ may accumulate anywhere on the limiting cusp-curve. Therefore, we must define the corresponding space $\mathcal{V}(D, T, J, \mathbf{z})$ to be the space $\mathcal{W}(D, T, J, p)$ considered in the previous section. In view of Theorem 5.3.1 (iii) (with N replaced by $N + 1$) this space has dimension

$$
\dim \mathcal{W}(D, T, J, p) = 2n + 2c_1(D) + 2p - 2N - 6
$$

$$\begin{aligned}
&\leq\; 2n + 2c_1(A) - 2(m-1)c_1(A^0) + 2p - 2N - 6 \\
&\leq\; 2n + 2c_1(A) - 2(m-1)c_1(A^0) + 4m - 2N - 6 \\
&=\; 2n + 2c_1(A) - 2(m-1)(c_1(A^0) - 2) - 2N - 2 \\
&\leq\; 2n + 2c_1(A) - 2 \\
&=\; \dim \mathcal{M}(A, J) - 2.
\end{aligned}$$

Here we have used $c_1(A^0) \geq 2$ which follows from (JA_p).

To complete the proof that these strata have the properties required by Theorem 5.4.1(iii), it remains to consider the special cases $p = 3, 4$ with the weakened assumptions stated there. Now the case $p = 3$ is covered by Theorem 5.3.1. Further, when $p = 4$ we must have $m \geq 2$. Hence we may assume that at least one of the components of the limit curve is multiply covered. Since, by (JA_4), every J-effective class B satisfies $c_1(B) \geq 1$ this implies that

$$c_1(D) + 1 \leq c_1(A).$$

If $N \geq 2$ we obtain $\dim \mathcal{W}(D, T, J, 4) < \dim \mathcal{M}(A, J)$ as required. The only other case is $N = 1$. But then $D = \{B\}$ where $A = mB$ and so it follows from (JA_4) that $c_1(B) \geq 2$. This implies

$$c_1(D) + 2 \leq c_1(A)$$

and with $N = 1$ we obtain again that $\dim \mathcal{W}(D, T, J, 4) < \dim \mathcal{M}(A, J)$.

This covers the case (c). The case (d) occurs when the sequence $u_\nu(z)$ converges to a constant J-holomorphic curve on the complement of a finite set X. This case can be incorporated in the above treatment of the cases (a) and (b). Just take the special case where $m = 1$ and $A^0 \in H_2(M, \mathbb{Z})$ is the zero homology class. It is a simple matter to check that in this case we may remove the curve v^0 and instead consider all tuples $(v^1, \ldots, v^N, w_{\ell+1}, \ldots, w_N)$ which satisfy

$$v^1(0) = \cdots = v^\ell(0)$$

and

$$v^\nu(0) = v^{j_\nu}(w_\nu), \qquad \ell + 1 \leq \nu \leq N.$$

The corresponding domain of the evaluation map has dimension $2n + 2c_1(D) - 2N$ and since $N \geq 1$ this proves the required inequality. Here we only need to assume that J is K-semi-positive for some $K > \omega(A)$ in order to ensure the inequality $c_1(D) \leq c_1(A)$. This completes the proof of Theorem 5.4.1.

Remark 6.6.3 The assumption that there are no J-holomorphic curves with negative Chern number plays a crucial role in the proof of the dimension formula in statement (iii) of all three Theorems 5.2.1, 5.3.1, and 5.4.1. If there is a J-holomorphic curve of class B with $c_1(B) < 0$ then an m-fold B-curve C^1 may appear in the limiting cusp-curve $C = C^1 \cup \ldots \cup C^N$. Then simplifying the curve C^1 would increase the dimension of the moduli space by $2(m-1)|c_1(B)|$. Hence the codimension argument will fail, unless one can prove that such multiply covered curves with negative Chern number cannot appear as components of limiting cusp-curves. So far no techniques have been developed to prove such statements for (nonintegrable) almost complex structures. Similar results will also be needed to remove the assumption (JA_p) in Theorem 5.4.1. The development of such results would constitute an important breakthrough in extending the symplectic invariants discussed in this paper (as well as Floer homology) to symplectic manifolds which are not weakly monotone. $\qquad\square$

Chapter 7

Gromov-Witten Invariants

The Gromov-Witten invariants count the number of isolated curves which intersect specified homology cycles in M. (Here a curve is called "isolated" if the dimension of the appropriate moduli space is zero.) In the case of Φ, one does not care where the curve intersects the cycles, while in defining Ψ these intersection points are fixed in the domain S^2.

Thus if $\mathcal{M}(A, J)$ denotes the moduli space of all J-holomorphic maps

$$u : \mathbb{C}P^1 \to M$$

which represent the homology class $A \in H_2(M; \mathbb{Z})$, consider the evaluation map

$$e_p : \mathcal{M}(A, J) \times_{\mathrm{G}} (\mathbb{C}P^1)^p \to M^p$$

given by

$$e_p(u, z_1, \ldots, z_p) = (u(z_1), \ldots, u(z_p))$$

which was discussed in Chapter 1. The invariant Φ counts the number of intersection points of the image of this map with a d-dimensional homology class α in M^p, where the dimension d is chosen so that, if all intersections were transverse, there would be a finite number of such points. To define Ψ, one takes a generic point $\mathbf{z} \in (\mathbb{C}P^1)^p$, where $p \geq 3$, and counts the number of points in which the image of $\mathcal{M}(A, J) \times \mathbf{z}$ meets the d-dimensional class α, where, again, d is chosen to make this number finite.

Since the moduli space $\mathcal{M}(A, J)$ is not in general compact, we must show that these numbers are well-defined, and independent of the choice of J and of representing cycle for α. To do this, it is convenient to introduce the notion of a pseudo-cycle. We shall see that the compactness theorems of Chapter 5 may be rephrased by saying that the evaluation map e_p, and its variant $e_{\mathbf{z}}$, represent pseudo-cycles in M^p. Then Φ and Ψ are simply the intersection numbers of these pseudo-cycles with homology classes of complementary dimension in M^p.

These invariants were first introduced into symplectic geometry in this generality by Ruan (cf. [64] and [65]). (Ruan denoted the invariant Ψ by $\tilde{\Phi}$) Special cases of the invariant Φ were used by Gromov in [26] and by McDuff in [46] in order to get information on the structure of symplectic manifolds. The invariant Ψ arises in the context of sigma models and was considered by Witten in [85]. It is also possible to define "mixed" invariants, in which one keeps track of the markings of only some of the points (cf. [67]). In the terminology of Kontsevich and Manin [35] the invariant Φ corresponds to the codimension-0 classes and the invariant Ψ to the classes of highest codimension while the mixed invariants constitute intermediate cases.

7.1 Pseudo-cycles

Let M be a smooth compact m-dimensional manifold. An arbitary subset $B \subset M$ is said to be of **dimension at most** m if it can be covered by the image of a map $g : W \to M$ which is defined on a manifold W of dimension m.[1] In this case we write $\dim B \leq m$. A k-dimensional **pseudo-cycle** in M is a smooth map

$$f : V \to M$$

defined on an oriented k-dimensional manifold V such that roughly speaking, the *boundary* of $f(V)$ is of dimension at most $k - 2$. More precisely, this *boundary* has to be defined as the set of all limit points of sequences $f(x_\nu)$ where x_ν has no convergent subsequence in V. This agrees with the notion of the **omega-limit-set** in dynamical systems, and we shall denote this set by

$$\Omega_f = \bigcap_{\substack{K \subset V \\ K \ compact}} \overline{f(V - K)}.$$

Note that this set is always compact. Note also that if V is the interior of a compact manifold $\bar{V}$ with boundary $\partial \bar{V} = \partial V$ and f extends to a continuous map $f : \bar{V} \to M$ then $\Omega_f = f(\partial V)$. The condition on f to be a pseudo-cycle is

$$\dim \Omega_f \leq \dim V - 2.$$

Two k-dimensional pseudo-cycles $f_0 : V_0 \to M$ and $f_1 : V_1 \to M$ are said to be **bordant** if there exists a $(k + 1)$-dimensional oriented manifold W with $\partial W = V_1 - V_0$ and a smooth map $F : W \to M$ such that

$$F|_{V_0} = f_0, \qquad F|_{V_1} = f_1, \qquad \dim \Omega_F \leq k - 1.$$

Pseudo-cycles in M form an abelian group with addition given by disjoint union. The neutral element is the empty map defined on the empty manifold $V = \emptyset$. The inverse of $f : V \to M$ is given by reversing the orientation of V. One could define pseudo-homology of M as the quotient group of pseudo-cycles modulo the bordism equivalence relation. It is not clear, however, whether the resulting homology groups agree with singular homology. We shall return to this question at the end of this section.

Remark 7.1.1 In order to represent a d-dimensional singular homology class α by a pseudo-cycle $f : V \to M$ represent it first by a map $f : P \to M$ defined on a d-dimensional finite oriented simplicial complex P without boundary. This condition means that the oriented faces of its top-dimensional simplices cancel each other out

---

[1] All our manifolds are σ-compact. This means that they can be covered by countably many compact sets.

in pairs. Thus P carries a fundamental homology class $[P]$ of dimension d and α is by definition the class $\alpha = f_*[P]$. Now approximate f by a map which is smooth on each simplex. Finally, consider the union of the d and $(d-1)$-dimensional faces of P as a smooth d-dimensional manifold V and approximate f by a map which is smooth across the $(d-1)$-dimensional simplices.

Of course, things become easier if we work with rational homology $H_*(M,\mathbb{Q})$. Because rational homology tensored with the oriented bordism group of a point is isomorphic to rational bordism $\Omega_*(M) \otimes \mathbb{Q}$, there is a basis of $H_*(M,\mathbb{Q})$ consisting of elements which are represented by smooth manifolds. Thus we may suppose that P is a smooth manifold, if we wish. $\qquad\square$

Exercise 7.1.2 Let M be compact smooth manifold of dimension dim $M \geq 3$. Prove that every 2-dimensional integral homology class $A \in H_2(M,\mathbb{Z})$ can be represented by an oriented embedded surface. **Hint:** Note first that A can be represented by a finite cycle, because M can be triangulated. Every such cycle can be thought of as a continuous map defined on a compact 2-dimensional simplicial complex without boundary. Every such complex can be given the structure of a smooth 2-dimensional compact manifold without boundary (which in the case of integer coefficients is orientable). Hence A is represented by a continuous map $f : \Sigma \to X$ defined on a smooth compact 2-manifold Σ. Approximate f by a smooth map and use a general position argument to make f an immersion with finitely many transverse self-intersections. In the cases dim $M = 3$ or dim $M = 4$ use a local surgery argument to remove the self-intersections. If dim $M > 4$ use a general position argument to obtain an embedding. $\qquad\square$

Two pseudo-cycles $e : U \to M$ and $f : V \to M$ are called **transverse** if

$$\Omega_e \cap \overline{f(V)} = \emptyset, \qquad \overline{e(U)} \cap \Omega_f = \emptyset,$$

and if

$$T_x M = \operatorname{im} de(u) + \operatorname{im} df(v)$$

whenever $e(u) = f(v) = x$. If e and f are transverse then the set

$$\{(u,v) \in U \times V \mid e(u) = f(v)\}$$

is a compact manifold of dimension dim U + dim V − dim M. In particular, this set is finite if U and V are of complementary dimension.

Lemma 7.1.3 *Let* $e : U \to M$ *and* $f : V \to M$ *be pseudo-cycles of complementary dimension.*

(i) *There exists a set* $\operatorname{Diff}_{\mathrm{reg}}(M,e,f) \subset \operatorname{Diff}(M)$ *of the second category such that* e *is transverse to* $\phi \circ f$ *for all* $\phi \in \operatorname{Diff}_{\mathrm{reg}}(M,e,f)$.

(ii) *If* e *is transverse to* f *then the set* $\{(u,v) \in U \times V \mid e(u) = f(v)\}$ *is finite. In this case define*

$$e \cdot f = \sum_{\substack{u \in U,\, v \in V \\ e(u)=f(v)}} \nu(u,v)$$

where $\nu(u,v)$ *is the intersection number of* $e(U)$ *and* $f(V)$ *at the point* $e(u) = f(v)$.

(iii) *The intersection number $e \cdot f$ depends only on the bordism classes of e and f.*

Proof: This is proved by standard arguments in differential topology as in [54] and [28]. Here are the main points. The first statement can be proved by the same techniques as the results of Chapter 6 with parameter space $\mathrm{Diff}(M)$ instead of $\mathcal{J}$. It is easy to see that the map

$$U \times V \times \mathrm{Diff}(M) \to M \times M : (u, v, \phi) \mapsto (e(u), \phi(f(v)))$$

is transverse to the diagonal and hence the *universal space*

$$\mathcal{M} = \{(u, v, \phi) \,|\, e(u) = \phi(f(v))\} \subset U \times V \times \mathrm{Diff}(M)\}$$

is a manifold. Now consider the regular values of the projection $\mathcal{M} \to \mathrm{Diff}(M)$ to obtain that e and $\phi \circ f$ are transverse for generic ϕ. Now choose smooth maps $e_0 : U_0 \to M$ and $f_0 : V_0 \to M$ such that $\dim U_0 = \dim U - 2$ and $\Omega_e \subset e_0(U_0)$ and similarly for f. Apply the same argument as above to the pairs (e_0, f), (e, f_0), (e_0, f_0) to conclude that $\overline{e(U)} \cap \phi(\Omega_f) = \emptyset$ and $\Omega_e \cap \overline{\phi(f(V))} = \emptyset$ for generic ϕ. This proves (i).

Statement (ii) is obvious. To prove statement (iii) assume that the pseudo-cycle $f : V \to M$ is bordant to the empty set with corresponding bordism $F : W \to M$. As in (i) it can be proved that this bordism can be chosen transverse to e and, moreover,

$$\Omega_e \cap \overline{F(W)} = \emptyset, \qquad \overline{e(U)} \cap \Omega_F = \emptyset.$$

Hence the set

$$X = \{(u, v) \in U \times W, \,|\, e(u) = F(v)\}$$

is a compact oriented 1-manifold with boundary

$$\partial X = \{(u, v) \in U \times V \,|\, e(u) = f(v)\}.$$

This proves that $e \cdot f = 0$. $\square$

Every $(m - d)$-dimensional pseudo-cycle $e : W \to M$ determines a homomorphism

$$\Phi_e : H_d(M, \mathbb{Z}) \to \mathbb{Z}$$

as follows. Represent the class $\beta \in H_d(M, \mathbb{Z})$ by a cycle $f : V \to M$ as in Remark 7.1.1. Any two such representations are bordant and hence, by Lemma 7.1.3, the intersection number

$$\Phi_e(\beta) = e \cdot f \tag{7.1}$$

is independent of the choice of the cycle f representing β. The next assertion also follows from Lemma 7.1.3.

Lemma 7.1.4 *The homomorphism Φ_e depends only on the bordism class of e.*

Thus we have proved that every $(m - d)$-dimensional pseudo-cycle $e : U \to M$ determines a homomorphism $\Phi_e : H_d(M, \mathbb{Z}) \to \mathbb{Z}$ and hence an element of the space $H^d(M) = H^d(M, \mathbb{Z})/\mathrm{torsion} = \mathrm{Hom}(H_d(M, \mathbb{Z}), \mathbb{Z})$ which we denote by

$$a_e = [\Phi_e] \in H^d(M).$$

Thus we have defined a map from bordism classes of pseudo-cycles to $H^d(M)$. If e is actually a smooth cycle (as in Remark 7.1.1) then the homology class of e is the Poincaré dual of $a_e = \mathrm{PD}([e])$. In general, one can think of e as a *weak representative* of the class $\alpha_e = \mathrm{PD}(a_e) \in H_{m-d}(M)$. In other words, an $(m-d)$-dimensional pseudo-cycle $e : U \to M$ is called a **weak representative** of the homology class $\alpha \in H_{m-d}(M)$ if $e \cdot f = \alpha \cdot \beta$ for every homology class $\beta \in H_d(M, \mathbb{Z})$ and every cycle f representing β (as in Remark 7.1.1). That this is a meaningful definition requires the proof that the formula (7.1) continues to hold when $f : V \to M$ is only a pseudo-cycle representing the class β in the weak sense just defined. This assertion is not obvious because we do not know whether any two weak representatives of a homology class $\beta \in H_d(M)$ are bordant.

Lemma 7.1.5 *Let $e : U \to M$ be an $(m-d)$-dimensional pseudo-cycle. If the d-dimensional pseudo-cycle $f : V \to M$ is a weak representative of the homology class $\beta \in H_d(M)$ then $\Phi_e(\beta) = e \cdot f$.*

Proof: It suffices to prove the assertion in the case $\beta = 0$. Hence assume that f has intersection number 0 with every $(m-d)$-dimensional smooth cycle. We must prove that f has intersection number 0 with every $(m-d)$-dimensional pseudo-cycle. To see this assume that the $(m-d)$-dimensional pseudo-cycle $e : U \to M$ is in general position. Thus $\overline{e(U)}$ does not intersect the (compact) limit set Ω_f. Now choose a sufficiently small open neighbourhood $W \subset M$ of Ω_f with smooth boundary, transverse to f. Then $V_0 = f^{-1}(M - W)$ is a compact manifold with boundary and the restriction of f to V_0 is a smooth map $f_0 : (V_0, \partial V_0) \to (M - W, \partial W)$. This map has intersection number zero with every $(m-d)$-dimensional cycle in $M - W$. Hence there is an integer $\ell > 0$ such that the ℓ-fold multiple of f_0 is a boundary in $H_d(M - W, \partial W; \mathbb{Z})$. This implies that the pseudo-cycle $e : U \to M - W$ has intersection number zero with f_0 and hence with f. $\qquad\square$

7.2 The invariant Φ

Let (M, ω) be a compact symplectic manifold and fix a class $A \in H_2(M, \mathbb{Z})$. We will assume throughout that (M, ω) and A satisfy the following hypotheses.

$(H1)$ *The class A is not a nontrivial multiple of a class B with $c_1(B) = 0$.*

$(H2)$ *The manifold (M, ω) is weakly monotone.*

Recall from Section 5.1 that a symplectic manifold (M, ω) is called weakly monotone if every spherical homology class $A \in H_2(M, \mathbb{Z})$ with $\omega(A) > 0$ and $c_1(A) < 0$ must satisfy $c_1(A) \le 2 - n$. This condition guarantees that, for generic J, there are no J-holomorphic spheres in homology classes of negative first Chern number. As we saw in Chapters 5 and 6, this is the essential condition which makes the compactness theorems work.

Given a spherical homology class $A \in H_2(M, \mathbb{Z})$ we shall define homomorphisms

$$\Phi_{A,p} : H_d(M^p, \mathbb{Z}) \to \mathbb{Z}$$

for $p \ge 1$ where

$$d = 2(n-1)(p-1) + 4 - 2c_1(A). \tag{7.2}$$

Roughly speaking, the number $\Phi_{A,p}(\alpha)$ counts (with multiplicities) the number of J-holomorphic curves in the homology class A whose p-fold product intersects a generic representative of the class $\alpha \in H_d(M^p, \mathbb{Z})$. The condition (7.2) guarantees that for a generic choice of the almost complex structure J there are finitely many such curves. Mostly we shall evaluate $\Phi_{A,p}$ on homology classes $\alpha \in H_d(M^p, \mathbb{Z})$ which are products of classes $\alpha_j \in H_{d_j}(M, \mathbb{Z})$ with $d_1 + \cdots + d_p = d$ and, for such α, will denote $\Phi_{A,p}(\alpha)$ by

$$\Phi_A(\alpha_1, \ldots, \alpha_p)$$

suppressing the notation p. Thus we may think of $\Phi_{A,p}$ as a collection of homomorphisms

$$H_{d_1}(M, \mathbb{Z}) \otimes \cdots \otimes H_{d_p}(M, \mathbb{Z}) \to \mathbb{Z}$$

where

$$d = d_1 + \cdots + d_p$$

satisfies (7.2).

To define $\Phi_{A,p}$ we fix a generic ω-tame almost complex structure $J \in \mathcal{J}(M, \omega)$. (More precisely, it should be regular in the sense of Theorem 5.3.1. See also Remark 5.1.1.) By Theorem 3.1.2, the space $\mathcal{M}(A, J)$ of J-holomorphic curves has dimension $2n + 2c_1(A)$. Further, formula (7.2) shows that the domain

$$\mathcal{W}(A, J, p) = \mathcal{M}(A, J) \times_{\mathrm{G}} (\mathbb{C}P^1)^p$$

of the p-fold evaluation map e_p has dimension

$$\begin{aligned}
\dim \mathcal{W}(A, J, p) &= 2n + 2c_1(A) + 2p - 6 \\
&= 2np - d \\
&= \sum_{i=1}^{p}(2n - d_i).
\end{aligned}$$

where $d_1 + \cdots + d_p = d$. Choose homology classes $\alpha_i \in H_{d_i}(M; \mathbb{Z})$ for $1 \le i \le p$. Then the set of A-curves which intersect the classes $\alpha_1, \ldots, \alpha_p$ is a finite set provided that the cycles which represent these classes are *in general position*. The invariant $\Phi_{A,p}(\alpha_1, \ldots, \alpha_p)$ is defined as the number of such curves, counted with appropriate signs:

$$\Phi_{A,p}(\alpha_1, \ldots, \alpha_p) = \#\left\{ [u, z_1, \ldots, z_p] \in \mathcal{W}(A, J, p) \,|\, u(z_i) \in \alpha_i \right\}.$$

(A more formal definition is given below.) We shall see that, in the weakly monotone case, this invariant is independent of the choice of J. Moreover, the form ω is not actually needed for the definition of the invariant $\Phi_{A,p}$ except as a tool to prove finiteness. It follows that $\Phi_{A,p}$ does not change when ω varies, provided that it remains weakly monotone.

Evaluation maps as pseudo-cycles

The assertion of Theorem 5.3.1 can be restated in the form that the evaluation map

$$e_p = e_{A,J,p} : \mathcal{W}(A, J, p) = \mathcal{M}(A, J) \times_{\mathrm{G}} (\mathbb{C}P^1)^p \to M^p$$

defined by $e_p(u, z_1, \ldots, z_p) = (u(z_1), \ldots, u(z_p))$ determines a pseudo-cycle whenever $J \in \mathcal{J}_+(M, \omega, K)$ for some $K > \omega(A)$ and J is regular. To see this observe that the lower dimensional strata are given by the evaluation maps

$$e_{D,T} : \mathcal{W}(D, T, J, p) \to M^p.$$

All these manifolds $\mathcal{W}(D, T, J, p)$ are σ-countable and have dimension at least two less than that of $\mathcal{W}(A, J, p)$. Moreover, the union of the sets $e_{D,T}(\mathcal{W}(D, T, J, p))$ cover the omega-limit-set $\Omega_{e_{A,J,p}}$. Hence the map $e_{A,J,p} : \mathcal{W}(A, J, p) \to M^p$ is a pseudo-cycle as claimed.

It follows from the above discussion that this map determines a homomorphism

$$\Phi_{A,J,p} : H_d(M^p, \mathbb{Z}) \to \mathbb{Z}$$

with

$$d = 2(n-1)(p-1) + 4 - 2c_1(A),$$

(provided that $J \in \mathcal{J}_+(M, \omega, K)$ for some $K > \omega(A)$ and J is regular). More explicitly, represent the homology class $\alpha \in H_d(M^p, \mathbb{Z})$ by a cycle $f : V \to M^p$ as in Remark 7.1.1. By Lemma 7.1.3 (i), put this map in general position so that it becomes transverse to $e_{A,J,p}$. By Lemma 7.1.3 (ii), this implies that the set $e_p(\mathcal{W}(A, J, p)) \cap f(V)$ is finite. Define

$$\Phi_{A,J,p}(\alpha) = e_{A,J,p} \cdot f$$

as the oriented intersection number. Lemma 7.1.3 (iii) implies that the right hand side depends only on the bordism class of f and hence only on the homology class α. Moreover, Lemma 7.1.5 shows that the formula $\Phi_{A,J,p}(\alpha) = e_{A,J,p} \cdot f$ continues to hold for every pseudo-cycle $f : V \to M^p$ which is a weak representative of the class α. By Lemma 7.1.4, the homomorphism $\Phi_{A,J,p} : H_d(M^p, \mathbb{Z}) \to \mathbb{Z}$ depends only on the bordism class of the evaluation map $e_{A,J,p}$.

The invariant $\Phi_{A,J,p}(\alpha)$ vanishes on all the torsion elements $\alpha \in H_d(M^p, \mathbb{Z})$. Since the free part $H_d(M^p, \mathbb{Z})/\text{Tor}$ is generated by product cycles $\alpha = (\alpha_1 \times \ldots \times \alpha_p)$, one often restricts $\Phi_{A,J,p}$ to such cycles and writes

$$\Phi_{A,J,p}(\alpha_1, \ldots, \alpha_p) = \Phi_{A,J,p}(\alpha_1 \times \cdots \times \alpha_p).$$

It is convenient to set $\Phi_{A,J,p}(\alpha_1, \ldots, \alpha_p) = 0$ when the dimension condition (7.2) is not satisfied.

Remark 7.2.1 It is interesting to note that the invariant $\Phi_{A,J,p}(\alpha_1, \ldots, \alpha_p)$ can be defined as the intersection number of $e_{A,J,p}$ with a product cycle $f_1 \times \cdots \times f_p$ where the $f_j : V_j \to M$ are pseudo-cycles representing the homology classes α_j, respectively. In other words, transversality can be achieved within the class of product cycles. This is a simple adaptation of Lemma 7.1.3 (i) to the case where f is a product cycle and ϕ is a product diffeomorphism. $\qquad\square$

Proposition 7.2.2 *Assume that (M, ω) is weakly monotone and A is not a nontrivial multiple of a class B with $c_1(B) = 0$. Then the homomorphism*

$$\Phi_{A,p} = \Phi_{A,J,p} : H_d(M^p, \mathbb{Z}) \to \mathbb{Z}$$

is independent of the choice of the regular ω-tame almost complex structure $J \in \mathcal{J}_{\mathrm{reg}}(M, \omega)$ used to define it.

Proof: To prove this we need a version of Theorem 5.3.1 which is valid for regular homotopies $\{J_\lambda\}$. That is, we need to know that, given two regular almost complex structures J_0 and J_1, any path joining them may be perturbed so that the sets

$$\mathcal{W}(D, T, \{J_\lambda\}_\lambda, p) = \bigcup_\lambda \{\lambda\} \times \mathcal{W}(D, T, J_\lambda, p)$$

are smooth manifolds for all D and T. This follows by the same argument which proves Theorem 5.3.1. We conclude that the evaluation maps

$$e_{A,J_0,p} : \mathcal{W}(A, J_0, p) \to M^p, \qquad e_{A,J_1,p} : \mathcal{W}(A, J_1, p) \to M^p,$$

determine bordant pseudo-cycles. The statement now follows from Lemma 7.1.4.
$\Box$

A deformation of a symplectic form ω is a smooth 1-parameter family of $\omega_t, t \in [0, 1]$ of forms starting at $\omega_0 = \omega$. In distinction to the notion of isotopy, we do not require that the cohomology class remain constant. Because the taming condition is open, an almost complex structure J which is tamed by ω is tamed by all sufficiently close symplectic forms. It follows easily that the homomorphism $\Phi_{A,p}$ does not change under a deformation ω_t of ω, provided that (M, ω_t) is weakly monotone for all t. Since manifolds of dimension ≤ 6 and manifolds where c_1 vanishes on spherical classes are always weakly monotone, $\Phi_{A,p}$ is a deformation invariant in these cases.

For part of our discussion the condition $(H2)$ can be replaced by the following weaker hypothesis. Again, the relevant definitions come from Section 5.1.

$(H3)$ *The set* $\mathcal{J}_+(M, \omega, K)$ *of all K-semi-positive almost complex structures which are compatible with ω is nonempty for some $K > \omega(A)$.*

If we replace the condition that (M, ω) be weakly monotone by the weaker hypothesis $(H3)$ then the homomorphism $\Phi_{A,J,p} : H_d(M^p, \mathbb{Z}) \to \mathbb{Z}$ can still be defined for generic $J \in \mathcal{J}_+(M, \omega, K)$ with $K > \omega(A)$. However, in this case the methods in the proof of Proposition 7.2.2 only show that $\Phi_{A,J,p}$ is locally independent of J and we do not know in general whether or not the space $\mathcal{J}_+(M, \omega, K)$ is path connected. Thus, to prove that they are invariant, i.e. independent of the almost complex structure J used to define them, we need hypothesis $(H2)$.

Another situation in which there is a well-defined invariant is when M is a Kähler manifold (M, J_0, ω) which satisfies one of the conditions mentioned in Remark 5.1.4. We formulate this here for the invariant Φ, but clearly a similar result holds for Ψ. Observe that here we somewhat change our perspective, thinking of the complex structure J_0 as given data rather than the symplectic form ω.

Proposition 7.2.3 *Suppose that (M, ω, J_0) is a Kähler manifold such that J_0 is tamed by a weakly monotone symplectic form ω'. Suppose also that A is not a nontrivial multiple of a class B with $c_1(B) = 0$. Then there is a neighbourhood $\mathcal{N}(J_0)$ of J_0 in the space of all almost complex structures on M such that the homomorphism*

$$\Phi_{A,p} = \Phi_{A,J,p} : H_d(M^p, \mathbb{Z}) \to \mathbb{Z}$$

is well-defined and independent of the choice of the regular element $J \in \mathcal{N}(J_0)$ used to define it, and of the taming form ω'. It depends on J_0 up to deformations through complex structures which are tamed by weakly monotone symplectic forms.

Proof: This is almost obvious. Since the taming condition is open, there is a path-connected neighbourhood $\mathcal{N}(J_0)$ of J_0 in the space of all almost complex structures on M, the elements of which are all tamed by ω'. Since ω' is monotone, the evaluation map $e_{A,J,p}$ will define a pseudo-cycle for generic $J \in \mathcal{N}(J_0)$. Just as before, its homotopy class is independent of the choice of the generic element $J \in \mathcal{N}(J_0)$, and of the choice of neighbourhood $\mathcal{N}(J_0)$. The rest of the proposition is clear. $\qquad\square$

Remark 7.2.4 (Integration over moduli spaces) There is an alternative way to express the invariant $\Phi_{A,p}$ in terms of the integrals of certain differential forms over the moduli space $\mathcal{W}(A, J, p)$. Namely, given a homology class $\alpha \in H_d(M^p)$ where d satisfies (7.2), we can define $\Phi_{A,J,p}(\alpha)$ as the integral

$$\Phi_{A,J,p}(\alpha) = \int_{\mathcal{W}(A,J,p)} e_{A,J,p}{}^* \tau.$$

Here the differential form τ is closed and represents the Poincaré dual $a = \mathrm{PD}(\alpha) \in H^{2n-d}(M^d)$. The dimension condition (7.2) asserts that

$$\deg \tau = 2n + 2c_1(A) + 2p - 6 = \dim \mathcal{W}(A, J, p)$$

and so $e_{A,J,p}{}^* \tau$ is a top-dimensional form on $\mathcal{W}(A, J, p)$. If one takes this approach one must show firstly that the integral is finite, secondly that it is independent of the differential form chosen to represent the class a, and thirdly as before that it is independent of the almost complex structure J. The first problem can be solved by choosing a differential form which is supported near the image of a generic pseudo-cycle $f : V \to M^p$ which represents the Poincaré dual $\alpha = \mathrm{PD}(a)$. If f is transverse to the evaluation map then the pullback $e_{A,J,p}{}^* \tau$ has compact support. This argument shows that the integral is finite for *some* form which represents the class a. To prove this for *all* forms representing a, and to prove that the integral is independent of the choice of the form, one must show that the pullback of any exact form integrates to zero, that is

$$\int_{\mathcal{W}(A,J,p)} e_{A,J,p}{}^* d\sigma = 0$$

for every form σ of degree $\dim \mathcal{W}(A, J, p) - 1$. Intuitively, this should be the case because the *boundary* of $\mathcal{W}(A, J, p)$ is of *codimension* 2 and so the integral of σ over it should vanish. However, to make this precise is somewhat nontrivial. In the case of intersection theory this problem can be avoided because of Sard's theorem. $\qquad\square$

In order to calculate $\Phi_{A,p}$, one has to be able to recognise when an almost complex structure J is generic (i.e. in the appropriate set $\mathcal{J}_{\mathrm{reg}}$). Our arguments show that J is generic if it is regular for all classes A^j which can appear as components of a reducible A-curve, and is such that all the relevant evaluation maps are transverse. For example, if J is regular for A-curves (which one can check by using Lemma 3.5.1) and if A is a simple class, then there are no reducible A-curves and J is generic in the required sense. In practice, one does not need all these conditions to be satisfied. All that is required to make the above constructions work is that J be "good", in the sense that J is regular for A-curves and that all the manifolds $\mathcal{W}(D, T, J, p)$ have dimension at least 2 less than that of $\mathcal{W}(A, J, p)$. Then the evaluation map $e_{A,J,p}$ does define a pseudo-cycle in the correct homology class.

Another important ingredient in the calculation of Φ_A is the determination of the correct signs. If J is integrable, we saw in Remark 3.3.6 that the moduli spaces have complex structures which are compatible with their orientations. Hence in this case it is usually easy to figure out the sign of intersection points (see Example 7.3.6 below). However, if J is not integrable, the question is much more delicate. When M has (real) dimension 4 and one is considering (non-empty) moduli spaces of spheres, McDuff in [47] shows that when $c_1(A) = 1 + p \geq 2$ the evaluation map

$$e_p : \mathcal{W}(A, J, p) \to M^p$$

preserves orientation. The condition $c_1(A) = 1 + p$ ensures that e_p maps between spaces of equal dimension, and so guarantees that $\Phi_{A,p}(\mathrm{pt}, \ldots, \mathrm{pt})$ is defined. Thus this result implies that whenever there is a J-holomorphic sphere with Chern number $1 + p \geq 2$ in a compact symplectic 4-manifold for any ω-tame J, the invariant

$$\Phi_{A,p}(\mathrm{pt}, \ldots, \mathrm{pt}) > 0.$$

It follows that the moduli space $\mathcal{M}(A, J)$ is never empty for any ω-tame J. This result has been proved only for curves of genus 0, and it is not clear what happens for a general symplectic 4-manifold when the genus is greater than 0. For example, the analogous result for tori would say that if there is a J-holomorphic torus with Chern number at least 1 for one ω-tame J, then there is such a torus for all ω-tame J. It is not known whether this holds. Some relevant examples can be found in Lorek [41].

7.3 Examples

Example 7.3.1 (Lines in Projective space) In $\mathbb{C}P^n$ two points lie on a unique line. This should translate into the statement that

$$\Phi_L(\mathrm{pt}, \mathrm{pt}) = 1,$$

where $L = [\mathbb{C}P^1] \in H_2(\mathbb{C}P^n, \mathbb{Z})$ and $\mathrm{pt} \in H_0(\mathbb{C}P^n, \mathbb{Z})$ is the homology class of a point.

To verify that this is indeed the case, note first that, since $c_1(A) = n + 1$ and $p = 2$, the dimension equation (7.2) works out correctly with $d = 0$. Further, the usual complex structure J_0 satisfies the conditions of Lemma 3.5.1. In fact, because $\mathbb{C}P^1 \subset \mathbb{C}P^2 \subset \mathbb{C}P^3$ and so on, the normal bundle to a complex line is a sum of line bundles each of Chern number 1. Therefore, J_0 satisfies the transversality requirement of Theorem 5.3.1 needed for the definition of Φ_L. (All the evaluation maps $e_D : \mathcal{M}(A_1, \ldots, A_N) \times (\mathbb{C}P^1)^{2N-2} \to M^{2N-2}$ are transverse to the diagonal

$\Delta_N \subset M^{2N-2}$.) Hence $\Phi_L(\mathrm{pt}, \mathrm{pt})$ is exactly the number of lines through two points, and so equals 1. $\qquad\square$

Example 7.3.2 (Conics in Projective space) In order to calculate the invariant $\Phi_A(\alpha_1, \ldots, \alpha_p)$ when $A = 2L$ using the standard complex structure J_0 one needs to check not only that J_0 is regular for the class $2A$ (which follows by Lemma 3.5.1) but also that the space of cusp-curves of type $D = (A, A)$ has small enough dimension. In fact, we need to check that the manifold $\mathcal{W}(D, J_0)$ which was defined in Chapter 5 above has dimension at least 2 less than $\dim \mathcal{W}(2L, J_0) = 6n$. This is an easy exercise. Thus, because there is a unique conic through 5 points in $\mathbb{C}P^2$, one finds that

$$\Phi_{2L}(\mathrm{pt}, \mathrm{pt}, \mathrm{pt}, \mathrm{pt}, \mathrm{pt}) = 1$$

in $\mathbb{C}P^2$. Further, if the α_i are all points then

$$\Phi_{2L}(\alpha_1, \ldots, \alpha_p) = 0$$

in $\mathbb{C}P^n$ with $n > 2$. To see this, observe that when $n = 3$ one would need $p = 4$ points to make condition (7.2) hold, but there is no conic through non-coplanar 4 points. (In fact, any three points determine a unique 2-plane P. If C is a conic through these three points then either $C \subset P$ or the intersection number $C \cdot P \geq 3$. But we know for homological reasons that $C \cdot P = 2$. Therefore C must be contained in P, which means that we cannot choose the 4th point freely.) Further, if $n > 3$ it is impossible to choose p so that condition (7.2) holds. Thus, if all the α_i are points, the invariants are rather limited. But with other choices of α_i one does get non-trivial invariants. For example, it is not hard to check that in $\mathbb{C}P^3$

$$\Phi_{2L}(\mathrm{pt}, \mathrm{pt}, \mathrm{pt}, \mathrm{line}, \mathrm{line}) = 1. \qquad\qquad\square$$

Example 7.3.3 (Cylinders) The result of Example 1.5.1 can be rephrased as the statement that

$$\Phi_A(\mathrm{pt}) = 1,$$

where $A = [\mathbb{C}P^1 \times \mathrm{pt}] \in H_2(\mathbb{C}P^1 \times V)$. By Proposition 7.2.2, this holds whenever $(\mathbb{C}P^1 \times V, \omega)$ is weakly monotone, and in particular if $\pi_2(V) = 0$ or $\dim V \leq 4$. $\square$

Example 7.3.4 (Rational and ruled symplectic 4-manifolds) Let (M, ω) be a compact symplectic 4-manifold which contains a symplectically embedded 2-sphere S with self-intersection number

$$S \cdot S \geq 0$$

but no such sphere with self-intersection number -1. Under these conditions it is proved in [45] (see also [51]) that there must be a symplectically embedded 2-sphere C of self-intersection number either 0 or 1. In the latter case the pair (M, C) is symplectomorphic to $(\mathbb{C}P^2, \mathbb{C}P^1)$. In the former case $C \cdot C = 0$ and M is diffeomorphic to a ruled surface, i.e. an S^2-bundle over a Riemann surface. The base is the moduli space of J-holomorphic spheres representing the homology class $A = [C]$ and the fibers are the J-holomorphic curves themselves (compare with the previous example). Positivity of intersection shows that these fibers are mutually disjoint and the adjunction formula for singular curves shows that they are embedded.

Example 7.3.5 (Exceptional divisors) Consider the manifold $-\mathbb{C}P^2$, which is the complex projective plane with the reversed orientation. If $A \in H_2(-\mathbb{C}P^2;\mathbb{Z})$ be the class of the complex line $\mathbb{C}P^1$ then $A \cdot A = -1$.

Now suppose that M is a complex Kähler surface, and let $\widetilde{M}$ denote its blow up at the point x. This means that the point x is replaced by the set of all lines through x. This set of lines is a copy of $\mathbb{C}P^1$ which is called the exceptional divisor Σ. The normal bundle to Σ is the canonical line bundle over $\mathbb{C}P^1$ which has Euler class -1, and it follows that $\widetilde{M}$ is diffeomorphic to the connected sum $M\#(-\mathbb{C}P^2)$ by a diffeomorphism which identifies Σ with a complex line in $-\mathbb{C}P^2$.[2] Let $E = [\Sigma]$ denote the homology class of the exceptional divisor. Then the moduli space $\mathcal{M}(E, J)$ consists of a single J-holomorphic curve, up to reparametrization, namely the exceptional divisor itself. To see this note that any other J-holomorphic curve in this class would have to intersect Σ with intersection number -1, but this is impossible because the intersection number of any two distinct J-holomorphic curves is nonnegative. Note that, by Lemma 3.5.2, the curve Σ is regular, and this is consistent with the dimension formula $\dim \mathcal{M}(E, J) = 4 + 2c_1(E) = 6$. Therefore we find that

$$\Phi_E(E) = E \cdot E = -1. \qquad\qquad \square$$

Example 7.3.6 (Non-deformation equivalent 6-manifolds) This example is due to Ruan (cf. [65]). Let X be $\mathbb{C}P^2$ with 8 points blown up, and Y be Barlow's surface. These manifolds are simply connected and have the same intersection form. Hence they are homeomorphic, but they are not diffeomorphic. By results of Wall, the stabilized manifolds $X' = X \times \mathbb{C}P^1$ and $Y' = Y \times \mathbb{C}P^1$ are diffeomorphic. Moreover, each of them is Kähler, and one can choose the diffeomorphism

$$\psi : X' \to Y'$$

so that it preserves the Chern classes of the respective complex structures. The latter statement implies that ψ takes the complex structure J_X on X' to a structure $\psi_*(J_X)$ which is homotopic to the obvious complex structure J_Y on Y'. Moreover an easy argument using cup-products shows that $\psi_*[X \times \text{pt}] = [Y \times \text{pt}]$, and this can be used to show that any class $A = [\Sigma \times \text{pt}] \in H_2(X')$ is mapped to a class of the form $\psi_* A = [\Sigma' \times \text{pt}] \in H_2(Y')$.

Now, on any Kähler manifold, the set of Kähler forms is a convex cone, and so any two such forms are deformation equivalent. (In fact, ω and ω' can be joined by the deformation $t\omega + (1 - t)\omega'$.) In other words, a Kähler manifold Z carries a natural deformation class of symplectic forms ω_Z. It is natural to ask whether the Kähler manifolds $(X', \omega_{X'})$ and $(Y', \omega_{Y'})$ are deformation equivalent. Ruan shows that they are not, by calculating

$$\Phi_E(E)$$

where $E \in H_2(X',\mathbb{Z})$ represents the class of one of the blown up points in X. Since $c_1(E) = p = 1$, the dimension requirement

$$2c_1(E) = 2(n - 1)(p - 1) + 4 - \dim \alpha,$$

is satisfied. Further, since there is exactly one holomorphic sphere Σ in X which represents E, the E-curves in $X' = X \times \mathbb{C}P^1$ are precisely the curves $\Sigma \times \{z\}$, for

---

[2]This is all fully explained in Griffiths and Harris [25], or McDuff and Salamon [52].

$z \in \mathbb{C}P^1$. Hence the set of points $X(E, J_E)$ which lie on J_X-holomorphic E-curves is just $\Sigma \times \mathbb{C}P^1$. Since $E \cdot E = -1$ in X, it is easy to check that the intersection number $E \cdot X(E, J_X) = -1$, and hence

$$\Phi_E(E) = -1.$$

However, the Barlow surface Y is minimal and so has no holomorphic curves with self-intersection number -1. Hence there are no J_Y-spheres in $Y' = Y \times \mathbb{C}P^1$ which represent the class $\psi_* E$. Here we use the fact that $\psi_* E = [\Sigma' \times \mathrm{pt}]$ and so any J_Y-holomorphic representative of $\psi_* E$ would have to be of the form $\Sigma' \times \mathrm{pt}$ where $\Sigma' \cdot \Sigma' = -1$. Thus J_Y is generic in the requisite sense, and so $\Phi_{\psi_* E}$ vanishes on Y'. But if ψ were a deformation equivalence then $\Phi_{\psi_* E}(\psi_* E)$ would have to be nonzero. $\square$

Example 7.3.7 (Counting discrete curves) If $2c_1(A) = 6 - 2n$ then the space $\mathcal{C}(A, J)$ of unparametrized A-curves is discrete (i.e. has dimension 0) and one might want to count the number $N(A)$ of these curves, with appropriate signs corresponding to the natural orientation of moduli space. These orientations can be defined as in Chapter 3. (It makes no difference that the moduli space is of dimension zero. Also, if J is integrable, it follows from Remark 3.3.6 that the orientation of each point is positive though it need not be for general J. See [42] and [64].) In order for condition $(H1)$ to hold, we need $c_1(A) > 0$, which restricts us to the case $2n = 4$. In this case there is, by Poincaré duality, another class $\alpha \in H_2(M)$ such that $A \cdot \alpha = k \neq 0$, and it is easy to check that

$$\Phi_A(\alpha) = kN(A).$$

Observe further, that if A is represented by an embedded sphere, the adjunction formula implies that $A \cdot A = -1$. As in Example 7.3.5, it then follows from positivity of intersections that $N(A) = 1$. However, this need not be true in general.

When $2n = 6$, there is a way to count the curves, avoiding the problems caused by multiple covers, which we describe in Section 9.3 below. When $2n > 6$ and the space $\mathcal{M}(A, J)$ is nonempty discrete curves must have $c_1(A) < 0$. Therefore condition $(H1)$ cannot hold and the present methods do not treat this case. $\square$

7.4 The invariant Ψ

We assume in this section that a generic almost complex structure J on M and the homology class $A \in H_2(M)$ satisfy the condition (JA_p) of Section 5.4. We will consider the following additional conditions.

$(H4)$ If $A = mB$ is a nontrivial multiple of a homology class B with $m > 1$ then either $c_1(B) \geq 3$ or $p \leq 2m$.

$(H5)$ For a generic almost complex structure $J \in \mathcal{J}(M, \omega)$ every J-effective homology class $A \in H_2(M)$ has Chern number $c_1(A) \geq 2$.

Note that in the cases $p = 3$ and $p = 4$ we always have $p \leq 2m$ and so condition $(H4)$ is empty. Condition $(H5)$ is satisfied, for example, if the manifold (M, ω) is weakly monotone with minimal Chern number $N \geq 2$. This implies that either (M, ω) is monotone or the minimal Chern number is $N \geq n - 2$. However, the important case of Calabi-Yau manifolds with Chern class $c_1 = 0$ is excluded by these assumptions.

If $(H4)$ and $(H5)$ are satisfied we shall define a homomorphism

$$\Psi_{A,p} : H_d(M^p, \mathbb{Z}) \to \mathbb{Z}$$

for $p \geq 1$ where

$$d = 2n(p - 1) - 2c_1(A). \tag{7.3}$$

Note that in the case $p = 3$ this number d agrees with the one given by (7.2). In fact for $p = 3$ the invariant $\Psi_{A,p}$ agrees with $\Phi_{A,p}$, but for $p > 3$ these invariants are different. Roughly speaking, we shall fix a p-tuple $\mathbf{z} = (z_1, \ldots, z_p) \in (\mathbb{C}P^1)^p$ and homology classes $\alpha_j \in H_{d_j}(M, \mathbb{Z})$ with

$$d_1 + \cdots + d_p = d,$$

and define

$$\Psi_{A,p}(\alpha_1, \ldots, \alpha_p) = \# \left\{ u \in \mathcal{M}(A, J) \,|\, u(z_j) \in \alpha_j \right\}.$$

The condition (7.3) will guarantee that, for a generic almost complex structure J and a generic point $\mathbf{z} \in (\mathbb{C}P^1)^p$, this is a finite set. The invariant $\Psi_{A,p}$ is the number of points in this set, counted with appropriate signs.

In order to make this precise we shall assume that J is regular in the sense of Definition 3.1.1 so that the moduli space $\mathcal{M}(A, J)$ is a manifold of dimension $2n + 2c_1(A)$. Now consider the evaluation map

$$e_{A,J,\mathbf{z}} : \mathcal{M}(A, J) \to M^p$$

defined by

$$e_{A,J,\mathbf{z}}(u) = (u(z_1), \ldots, u(z_p)).$$

Theorem 5.4.1 shows that for generic J the closure of the image of this map can be represented as the union of $e_{\mathbf{z}}(\mathcal{M}(A, J))$ with lower dimensional strata. In other words the map $e_{A,J,\mathbf{z}}$ represents again a pseudo-cycle and therefore determines a homomorphism

$$\Psi_{A,J,\mathbf{z}} : H_d(M^p, \mathbb{Z}) \to \mathbb{Z}, \qquad d = 2n(p - 1) - 2c_1(A).$$

As before, represent a homology class

$$\alpha = \alpha_1 \times \cdots \times \alpha_p \in H_d(M^p, \mathbb{Z})$$

by a pseudo-cycle $f = f_1 \times \cdots \times f_p : V_1 \times \cdots \times V_p \to M^p$ (see Remark 7.1.1). By Lemma 7.1.3 (i) put this map in general position so that it becomes transverse to the maps $e_{D,T,\mathbf{z}}$ for all D and T. By Lemma 7.1.3 (ii) this implies that the set $e_{A,J,\mathbf{z}}(\mathcal{M}(A, J)) \cap f(V)$ is finite. Thus, we may define

$$\Psi_{A,J,\mathbf{z}}(\alpha_1, \ldots, \alpha_p) = e_{A,J,\mathbf{z}} \cdot f$$

as the oriented intersection number. By Lemma 7.1.3 (iii) the right hand side depends only on the bordism class of f and hence only on the homology class α. Further, by Lemma 7.1.4, this map depends only on the bordism class of the evaluation map $e_{A,J,\mathbf{z}}$.

Lemma 7.4.1 *Assume* $(H4)$ *and* $(H5)$. *Then the homomorphism*

$$\Psi_{A,p} = \Psi_{A,J,\mathbf{z}} : H_d(M^p, \mathbb{Z}) \to \mathbb{Z}$$

is independent of the choice of the regular ω-tame almost complex structure $J \in \mathcal{J}_{\mathrm{reg}}(M, \omega)$ and the p-tuple $\mathbf{z} = (z_1, \ldots, z_p) \in (\mathbb{C}P^1)^p$ used to define it. Hence $\Psi_{A,p}$ depends only on the deformation class of ω.

Proof: As in the case of Proposition 7.2.2 the proof relies on a version of Theorem 5.4.1 for regular homotopies $\{J_\lambda\}$ and $\{\mathbf{z}_\lambda\}$. We must show that any two regular almost complex structures J_0 and J_1 and any two points $\mathbf{z}_0, \mathbf{z}_1 \in (\mathbb{C}P^1)^p$ can be joined by regular paths $\{J_\lambda\}$ and $\{\mathbf{z}_\lambda\}$ such that the spaces

$$\mathcal{V}(D, T, \{J_\lambda\}, \{\mathbf{z}_\lambda\}) = \bigcup_\lambda \{\lambda\} \times \mathcal{V}(D, T, J_\lambda, \mathbf{z}_\lambda)$$

are smooth manifolds for all D and T. This follows by the same arguments used in the proof of Theorem 5.4.1. We conclude that the evaluation maps

$$e_{A,J_0,\mathbf{z}_0} : \mathcal{M}(A, J_0) \to M^p, \qquad e_{A,J_1,\mathbf{z}_1} : \mathcal{M}(A, J_1) \to M^p,$$

determine bordant pseudo-cycles. The statement now follows from Lemma 7.1.4. $\square$

Remark 7.4.2 (Integration over moduli spaces) As in Remark 7.2.4 the invariant $\Psi_{A,J,\mathbf{z}}(\alpha_1, \ldots, \alpha_p)$ can be represented as an integral of a differential form over the moduli space $\mathcal{M}(A, J)$. Represent the cohomology classes $a_j = \mathrm{PD}(\alpha_j) \in H^{2n-d_j}(M)$ by closed forms $\tau_j \in \Omega^{2n-d_j}(M)$ and define

$$\Psi_{A,J,\mathbf{z}}(\alpha_1, \ldots, \alpha_p) = \int_{\mathcal{M}(A,J)} e_1^* \tau_1 \wedge e_2^* \tau_2 \wedge \ldots \wedge e_p^* \tau_p$$

where $e_j : \mathcal{M}(A, J) \to M$ denotes the evaluation map $e_j(u) = u(z_j)$. The dimension condition (7.3) means that

$$\sum_{j=1}^{p} \deg \tau_j = 2n + 2c_1(A) = \dim \mathcal{M}(A, J)$$

and so the exterior product of the forms $e_j^* \tau_j$ is a top-dimensional form on $\mathcal{M}(A, J)$. The difficulties in making this approach rigorous are the same as those discussed in Remark 7.2.4. $\square$

Remark 7.4.3 Both invariants Φ_A and Ψ_A as well as the mixed invariants of Ruan and Tian [67] can be viewed as special cases of the Gromov-Witten classes as formulated by Kontsevich and Manin [35]. More precisely, let Σ be a compact oriented Riemann surface of genus g and denote by $\mathcal{J}(\Sigma)$ the space of complex structures on Σ which are compatible with the given orientation. A $(k+1)$-tuple $(j, z_1, \ldots, z_k) \subset \mathcal{J}(\Sigma) \times \Sigma^k$ is called **stable** if the only diffeomorphism $\phi \in \mathrm{Diff}(\Sigma)$ which satisfies $\phi^* j = j$ and $\phi(z_i) = z_i$ for $i = 1, \ldots, k$ is the identity. Let $\mathcal{C}_{g,k} \subset$

$\mathcal{J}(\Sigma) \times \Sigma^k$ denote the space of such stable $(k+1)$-tuples and consider the quotient space

$$\mathcal{M}_{g,k} = \frac{\mathcal{C}_{g,k}}{\mathrm{Diff}(\Sigma)} \subset \frac{\mathcal{J}(\Sigma) \times \Sigma^k}{\mathrm{Diff}(\Sigma)}.$$

This is a manifold of dimension $\dim \mathcal{M}_{g,k} = 6g - 6 + 2k$. Now consider the space $\mathcal{C}_{g,k}(M, A)$ of all $(k+2)$-tuples $(u, j, z_1, \ldots, z_k)$ where $(j, z_1, \ldots, z_k) \in \mathcal{C}_{g,k}$ and $u : \Sigma \to M$ is a simple (j, J)-holomorphic curve representing the class A. The group $\mathrm{Diff}(\Sigma)$ acts on this space by

$$\phi^*(u, j, z_1, \ldots, z_k) = (u \circ \phi, \phi^* j, \phi^{-1}(z_1), \ldots, \phi^{-1}(z_k))$$

and, for a generic $J \in \mathcal{J}(M, \omega)$, the quotient space

$$\mathcal{M}_{g,k}(M, A) = \frac{\mathcal{C}_{g,k}(M, A)}{\mathrm{Diff}(\Sigma)} \subset \frac{\mathrm{Map}(\Sigma, M) \times \mathcal{J}(\Sigma) \times \Sigma^k}{\mathrm{Diff}(\Sigma)}$$

is a manifold of dimension $\dim \mathcal{M}_{g,k}(M, A) = (n - 3)(2 - 2g) + 2c_1(A) + 2k$. This gives rise to a linear map

$$\mathrm{GW}_{A,g,k} : H^*(M^k) \to H^*(\mathcal{M}_{g,k})$$

defined, heuristically, by

$$\mathrm{GW}_{A,g,k}(a_1, \ldots, a_k) = \mathrm{PD}(\pi_* \mathrm{PD}(e_1^* a_1 \wedge \ldots \wedge e_k^* a_k))$$

for $a_i \in H^*(M)$ where $e_i : \mathcal{M}_{g,k}(M, A) \to M$ is the obvious evaluation map and $\pi : \mathcal{M}_{g,k}(M, A) \to \mathcal{M}_{g,k}$ is the obvious projection. We shall not deal here with the rigorous definition of these invariants for higher genus which involves suitable compactifications of all the relevant moduli spaces. (A rigorous treatment of these higher genus invariants will appear in the forthcoming paper [70] of Ruan-Tian.) Note that

$$\deg \mathrm{GW}_{A,g,k}(a_1, \ldots, a_k) = n(2 - 2g) + 2c_1(A) - \sum_{i=1}^{k} \deg a_i.$$

The invariant Ψ corresponds to the case where this degree is zero and is given by evaluating the cohomology class $\mathrm{GW}_{A,g,k}(a_1, \ldots, a_k)$ at a point. The invariant Φ on the other hand corresponds to the case where this degree agrees with the dimension of the space $\mathcal{M}_{g,k}$ and is given by evaluating the cohomology class $\mathrm{GW}_{A,g,k}(a_1, \ldots, a_k)$ on the fundamental cycle $[\mathcal{M}_{g,k}]$. This is what Kontsevich and Manin call a *codimension-0 class*. The mixed invariants of Ruan and Tian [69] in the case $g = 0$ and $k \geq 3$ are given by evaluating a class $\mathrm{GW}_{A,0,k}(a_1, \ldots, a_k)$ of degree $2d$ over the $2d$-dimensional cycle $K(w_1, \ldots, w_{k-d-3}) \subset \mathcal{M}_{0,k}$ of all those $[j, z_1, \ldots, z_k] \in \mathcal{M}_{0,k}$ with $z_i = w_i$ for $i \leq k - d - 3$. $\square$

Example 7.4.4 As an illustration of the difference between Φ and Ψ consider conics in the complex projective plane. As we remarked in Section 7.3, the fact that there is a unique conic through 5 points implies that

$$\Phi_{2L}(\mathrm{pt}, \mathrm{pt}, \mathrm{pt}, \mathrm{pt}, \mathrm{pt}) = 1.$$

However, when $A = 2L$ and all the α_i are points, the dimensional condition (7.3) is satisfied when $p = 4$ not $p = 5$. Thus

$$\Psi_{2L,p}(\mathrm{pt}, \ldots, \mathrm{pt}) = 0$$

unless $p = 4$. It is easy to check that there is exactly one map of degree 2 from $\mathbb{C}P^1 \to \mathbb{C}P^2$ such that

$$[1:0] \mapsto [1:0:0], \quad [0:1] \mapsto [0:1:0],$$

$$[1:1] \mapsto [0:0:1], \quad [1:x] \mapsto [1:1:1],$$

namely $[s:t] \mapsto [xs(s-t):t(s-t):(1-x)st]$. Thus

$$\Psi_{2L}(\mathrm{pt}, \mathrm{pt}, \mathrm{pt}, \mathrm{pt}) = 1.$$

To understand this, note that the pair $(\mathbb{C}P^1, \mathbf{z})$ may be considered as a marked sphere. When $p > 3$, these markings have moduli (i.e. they are not all equivalent under the reparametrization group G). For example, if $p = 4$, the cross-ratio is the unique invariant of $\mathbf{z}$ modulo the action of G. Thus, the family of all conics which go through 4 fixed (and generic) points $q_1, \ldots, q_4$ in $\mathbb{C}P^2$ has real dimension 2. The above calculation shows that for each generic marking $\mathbf{z} = \{z_1, \ldots, z_4\}$, only one conic in this family may be parametrized so that each z_i is taken to the corresponding q_i. $\qquad\square$

In Section 9.1 we will show how to define Ψ_A in the weakly monotone case even for classes A with $c_1(A) = 0$. Note finally that in [66] Ruan defines invariants which count numbers of J-holomorphic tori, and uses them to obtain some information about the symplectic topology of elliptic surfaces.

Chapter 8

Quantum Cohomology

Much of the recent interest in holomorphic spheres and Gromov-Witten invariants has arisen because they may be used to define a new multiplication on the cohomology ring of a compact symplectic manifold. In the remaining chapters of this book, we shall explain this construction, give an outline of some of the proofs, and point out some of the relations to other subjects such as algebraic geometry, integrable Hamiltonian systems, and Floer homology. For more details, see for example Vafa [82], Aspinwall and Morrison [2], Ruan and Tian [67] and Givental and Kim [24].

We will begin in this chapter by assuming that M is monotone, i.e. that the cohomology class of ω is a positive multiple of c_1. In this context, quantum cohomology can be set up with coefficients in a polynomial ring. In the first two sections, we define the deformed cup product and explain why it should be associative. (The proof of associativity is given in Appendix A.) We then describe the Givental–Kim calculation of the quantum cohomology of flag manifolds and its relation to the Toda lattice. The chapter ends with a discussion of the Gromov-Witten potential and Dubrovin connection.

8.1   Witten's deformed cohomology ring

Triple intersections

We shall begin by reviewing the ordinary cup product on singular cohomology. To avoid difficulties with torsion we shall consider integral deRham cohomology $H^*(M) = H^*_{\mathrm{DR}}(M, \mathbb{Z})$, i.e. deRham cohomology classes which take integral values over all cycles. Thus we can identify $H^k(M)$ with $\mathrm{Hom}(H_k(M, \mathbb{Z}), \mathbb{Z})$.[1] Likewise, we denote by $H_*(M)$ the free part of $H_*(M, \mathbb{Z})$ or, more precisely, the quotient of $H_*(M, \mathbb{Z})$ by its torsion subgroup. Denote by

$$a(\beta) = \int_\beta a$$

---

[1] Of course, every cohomology class $a \in H^k(M, \mathbb{Z})$ determines a homomorphism $\phi_a :$ $H_k(M, \mathbb{Z}) \to \mathbb{Z}$ but this homomorphism may be zero for a nonzero cohomology class. The torsion in $H^*(M, \mathbb{Z})$ is, by definition, the kernel of the homomorphism $a \mapsto \phi_a$. In our notation $H^k(M)$ is isomorphic to the quotient of $H^k(M, \mathbb{Z})$ divided by the torsion subgroup.

the pairing of $a \in H^k(M)$ with $\beta \in H_k(M)$ and by $\alpha \cdot \beta$ the intersection pairing of two homology classes $\alpha \in H_{2n-k}(M)$ and $\beta \in H_k(M)$ of complementary dimension. This pairing determines a homomorphism $H_{2n-k}(M) \to H^k(M)$ which assigns to $\alpha \in H_{2n-k}(M)$ the cohomology class $a = \mathrm{PD}(\alpha) \in H^k(M)$ defined by

$$\int_\beta a = \alpha \cdot \beta$$

for $\beta \in H_k(M)$. Poincaré duality asserts that the map $\mathrm{PD} : H_{2n-k}(M) \to H^k(M)$ is an isomorphism. We shall denote its inverse also by PD. Now fix two cohomology classes $a \in H^k(M)$ and $b \in H^\ell(M)$ and denote by $\alpha \in H_{2n-k}(M)$ and $\beta \in H_{2n-\ell}(M)$ their respective Poincaré duals. Then the cup product $a \cup b \in H^{k+\ell}(M)$ is defined by the triple intersection

$$\int_\gamma a \cup b = \alpha \cdot \beta \cdot \gamma$$

for $\gamma \in H_{k+\ell}(M)$. Note that two cycles in general position representing α and β intersect in a pseudo-cycle of codimension $k + \ell$ and so the triple intersection is well defined. If $a, b \in H^*(M)$ are of complementary dimension then we shall use the notation

$$\langle a, b \rangle = a \cdot b = \int_M a \cup b.$$

In view of the above triple intersection formula this expression agrees with the intersection number $\alpha \cdot \beta$ of the corresponding Poincaré duals $\alpha = \mathrm{PD}(a)$ and $\beta = \mathrm{PD}(b)$.

Later on, we will consider elements $a \in \sum_i H^i(M)$ which are are sums $a = \sum_i a_i$ of elements $a_i \in H^i(M)$ of pure degree. It will then be convenient to extend the definition of the pairing $\langle a, b \rangle$, setting it equal to

$$\left\langle \sum_i a_i, \sum_j b_j \right\rangle = \sum_{i,j} \langle a_i, b_j \rangle,$$

where $\langle a_i, b_j \rangle = 0$ unless $i + j = 2n$.

Quantum cohomology

For now, we will suppose that (M, ω) is a $2n$-dimensional monotone symplectic manifold with minimal Chern number $N \geq 2$. Then, by Section 7.4, the invariant Ψ_A is defined for $p = 3$ and $p = 4$ and all homology classes A. Rescaling the form ω, if necessary, we may assume that $[\omega]$ is an integral class with $\omega(\pi_2(M)) = \mathbb{Z}$.

As explained in Chapter 1, there are several possible choices of coefficient ring for quantum cohomology. We will begin by defining quantum cohomology as the tensor product with the ring of Laurent polynomials

$$QH^*(M) = H^*(M) \otimes \mathbb{Z}[q, q^{-1}].$$

Here q is a variable of degree $2N$ and so the elements in $QH^*(M)$ of degree k can be expressed as finite sums

$$a = \sum_{i \in \mathbb{Z}} a_i q^i, \qquad a_i \in H^{k-2Ni}(M).$$

We shall denote by $QH^k(M)$ the set of elements of degree k. There is a natural Poincaré duality pairing $QH^*(M) \otimes QH^*(M) \to \mathbb{Z} : (a,b) \mapsto \langle a, b \rangle$ defined by

$$\langle a, b \rangle = \sum_{2N(i+j)=k+\ell-2n} a_i \cdot b_j \tag{8.1}$$

for $a = \sum_i a_i q^i \in QH^k(M)$ and $b = \sum_j b_j q^j \in QH^\ell(M)$. The condition $2N(i+j) = k+\ell-2n$ guarantees that a_i and b_j are of complementary dimension. We could also sum over all pairs (i,j) and simply define $a_i \cdot b_j = 0$ whenever $\deg(a_i)+\deg(b_j) \neq 2n$. Note that $\langle a, b \rangle = 0$ unless $\deg(a) + \deg(b) \equiv 2n (\text{mod } 2N)$. Note also that the pairing (8.1) is nondegenerate in the sense that $\langle a, b \rangle = 0$ for all b implies $a = 0$. However, it need not be positive definite. It is skew commutative in the sense that

$$\langle b, a \rangle = (-1)^{\deg(a)\deg(b)} \langle a, b \rangle.$$

for $a, b \in QH^*(M)$.

Remark 8.1.1 (i) Multiplication by q gives a natural isomorphism

$$QH^k(M) \cong QH^{k+2N}(M).$$

Moreover, there is an obvious isomorphism

$$QH^k(M) \cong \bigoplus_{j \equiv k (\text{mod } 2N)} H^j(M)$$

and hence the quantum cohomology of M can be thought of as the universal cover of the ordinary cohomology with a $\mathbb{Z}_{2N}$ grading.

(ii) We could also define $QH^*(M)$ as the tensor product with the polynomial ring $\mathbb{Z}[q]$. In fact we shall use the notation

$$\widetilde{QH}^*(M) = H^*(M) \otimes \mathbb{Z}[q]$$

for this alternative definition. In this case the elements of $\widetilde{QH}^k(M)$ are formal polynomials

$$a = \sum_{i \geq 0} a_i q^i$$

where again $a_i \in H^{k-2Ni}(M)$. On the first glance this may seem slightly simpler than the above definition. However, in this case the groups $\widetilde{QH}^k(M)$ vanish for $k \leq 0$ and become periodic only for $k \geq 2n$. We may in fact think of $\widetilde{QH}^*(M)$ as a subset of $QH^*(M)$. The restriction of the Poincaré duality pairing $(a,b) \mapsto \langle a, b \rangle$ to this subset is given by

$$\langle a, b \rangle = a_0 \cdot b_0$$

and is therefore degenerate.

(iii) As we shall see when we discuss flag manifolds in Section 8.3, it is sometimes possible to use the coefficient ring $\mathbb{Z}[q_1, \ldots, q_n]$ where n is the rank of $H_2(M)$ and the variables q_i can be thought of as (multiplicative) generators of $H_2(M)$. With this choice of coefficients the quantum cohomology groups, with the ring structure defined below, will carry more information. Roughly speaking, this is because J-holomorphic curves in different homology classes are counted separately, while, with our simplified definition, the information about the homology class of the curve is suppressed.

$\square$

Deformed cup product

The deformed cup product is a homomorphism

$$QH^\ell(M) \times QH^m(M) \to QH^{\ell+m}(M).$$

Because the Poincaré duality pairing is nondegenerate we can define the quantum deformed cup product $a * b$ of two classes

$$a = \sum_i a_i q^i \in QH^\ell(M), \qquad b = \sum_j b_j q^j \in QH^m(M)$$

in terms of their inner product with a third class

$$c = \sum_k c_k q^k \in QH^{2n-\ell-m}(M).$$

Recall that $a_i \in H^{\ell-2Ni}(M)$, $b_j \in H^{m-2Nj}(M)$, $c_k \in H^{2n-\ell-m-2Nk}(M)$ and denote their respective Poincaré duals by $\alpha_i = \mathrm{PD}(a_i)$, $\beta_j = \mathrm{PD}(b_j)$, and $\gamma_k = \mathrm{PD}(c_k)$. Then we define $a * b \in QH^{\ell+m}(M)$ by the formula

$$\langle a * b, c \rangle = \sum_{i,j,k} \sum_A \Phi_A(\alpha_i, \beta_j, \gamma_k) \tag{8.2}$$

for $c \in QH^*(M)$. Here the last sum runs over all classes $A \in H_2(M)$ which satisfy $N(i + j + k) + c_1(A) = 0$. This is precisely the condition needed in order for the expression $\Phi_A(\alpha_i, \beta_j, \gamma_k)$ to be defined. In fact the codimension of the class α_i (or rather of a pseudo-cycle representing α_i) is the degree of a_i and similarly for β_j and γ_k. These degrees satisfy the condition

$$\deg(a_i) + \deg(b_j) + \deg(c_k) = 2n + 2c_1(A). \tag{8.3}$$

This corresponds precisely to the dimension conditions (7.2) and (7.3) for the definition of the invariants Φ_A and Ψ_A. (Recall that these agree in the case $p = 3$.) The formula (8.3) shows that $-2n \leq c_1(A) \leq 2n$ and hence only finitely many values of $c_1(A)$ occur in (8.2). Because M is monotone, we have $\omega(A) = \lambda c_1(A)$ and hence the classes A which contribute nontrivially to the sum have uniformly bounded energy. Thus, by Corollary 4.4.4, only finitely many such A occur and hence the right hand side of (8.2) is a finite sum.

Remark 8.1.2 As discussed in chapter 1 the formula (8.2) can also be expressed as follows. Given ordinary cohomology classes $a \in H^\ell(M)$ and $b \in H^m(M)$ we define

$$a * b = \sum_d (a * b)_A \, q^{c_1(A)/N} \in QH^{\ell+m}(M)$$

by evaluating $(a * b)_A \in H^{\ell+m-2c_1(A)}(M)$ on a class $\gamma \in H_{\ell+m-2c_1(A)}(M)$:

$$\int_\gamma (a * b)_A = \Phi_A(\alpha, \beta, \gamma). \tag{8.4}$$

Here $\alpha = \mathrm{PD}(a) \in H_{2n-\ell}(M)$, $\beta = \mathrm{PD}(b) \in H_{2n-m}(M)$, and hence the classes α, β, γ satisfy the dimension condition

$$\deg(\alpha) + \deg(\beta) + \deg(\gamma) = 4n - 2c_1(A).$$

We leave it to the reader to check that this definition of $a * b$ is equivalent to (8.2). $\square$

Remark 8.1.3 The formula (8.2) involves only J-holomorphic curves with Chern number $0 \le c_1(A) \le 2n$. In Section 8.5 we shall discuss a family of deformed cup-products $(x, y) \mapsto x *_a y$ which are parametrized by cohomology classes $a \in H^*(M, \mathbb{C})$. The definition of these cup-products involves the invariants $\Phi_{A,p}$ for all $p \ge 3$ and hence J-holomorphic curves of all possible Chern numbers. The above construction corresponds to the case $a = 0$ and $p = 3$. $\square$

The constant term in the expansion (8.2) comes from counting A-curves with $A = 0$. Since J is ω-tame, ω restricts to an area form on any non-constant J-holomorphic curve. Thus the curves with $\omega(A) = 0$ are constant and $\Phi_0(\alpha, \beta, \gamma)$ is just the usual triple intersection index

$$\Phi_0(\alpha, \beta, \gamma) = \alpha \cdot \beta \cdot \gamma.$$

It follows that the constant term in the expansion of $a * b$ is just the ordinary cup product $a \cup b$. Thus this multiplication is a deformation of the cup product, as advertised.

Proposition 8.1.4 (i) *The quantum cup product is distributive over addition and skew-commutative in the sense that*

$$a * b = (-1)^{\deg(a)\,\deg(b)} b * a$$

for $a, b \in QH^*(M)$. It also commutes with the action of $\mathbb{Z}[q, q^{-1}]$.

(ii) *If $a \in H^0(M)$ or $a \in H^1(M)$ then the deformed cup-product agrees with the ordinary cup-product, i.e.*

$$a * b = a \cup b$$

for all $b \in H^*(M)$.

(iii) *The canonical generator $\mathbb{1} \in H^0(M)$ is the unit element in quantum cohomology.*

(iv) *The scalar product is invariant with respect to the cup product in the sense that*

$$\langle a * b, c \rangle = \langle a, b * c \rangle$$

for $a, b, c \in QH^*(M)$.

Proof: The first statement is obvious from the definition. To prove the second statement for $a \in H^0(M)$ or $a \in H^1(M)$ we must prove

$$\Phi_A(\alpha, \beta, \gamma) = 0$$

whenever $c_1(A) \neq 0$. We shall prove in fact that no A-curve will intersect both classes β and γ when these are in general position. The double evaluation map is defined on the space $\mathcal{W}(A, J, 2)$ of dimension

$$\dim \mathcal{W}(A, J, 2) = 2n + 2c_1(A) - 2.$$

Since $k = 0$ or $k = 1$ we have

$$2n + 2c_1(A) = \deg(a) + \deg(b) + \deg(c) \leq 1 + \deg(b) + \deg(c)$$

and hence

$$\dim \mathcal{W}(A, J, 2) \leq \deg(b) + \deg(c) - 1.$$

But the codimension of $\beta \times \gamma \subset M^2$ is $\deg(b) + \deg(c)$ and is therefore larger than the dimension of the domain $\mathcal{W}(A, J, 2)$ of the double evaluation map. Hence the cycle $\beta \times \gamma$ will not intersect the image of $e : \mathcal{W}(A, J, 2) \to M^2$ when in general position. This proves that the only nontrivial contribution to the deformed cup-product occurs when $c_1(A) = 0$ and this is given by

$$\int_\gamma (a * b)_0 = \alpha \cdot \beta \cdot \gamma = \int_\gamma a \cup b$$

This proves (ii), and (iii) follows immediately from (ii). To prove (iv) just note that $\phi_A(\alpha_i, \beta_j, \gamma_k)$ is skew symmetric under permutations of the three entries. In particular,

$$\Phi_A(\alpha_i, \beta_j, \gamma_k) = (-1)^{\deg(a)(\deg(b)+\deg(c))} \Phi_A(\beta_j, \gamma_k, \alpha_i)$$

and hence, taking the sum over all quadruples (i, j, k, A) with $N(i+j+k) + c_1(A) = 0$, we obtain

$$\langle a * b, c \rangle = (-1)^{\deg(a)(\deg(b)+\deg(c))} \langle b * c, a \rangle = \langle a, b * c \rangle$$

as claimed. $\square$

Remark 8.1.5 A similar argument as in the proof of the previous theorem shows that $\Phi_A(\alpha_1, \ldots, \alpha_p) = 0$ whenever one of the classes α_j has degree $\deg(\alpha_j) \geq 2n-1$.
$\square$

Associativity of the deformed cup-product is far from obvious and we defer this to Section 8.2. If this is proved then, together with Proposition 8.1.4, it shows that the quantum cohomology ring $QH^*(M)$ is a **Frobenius algebra**. The axioms for a Frobenius algebra are precisely the assertions of Proposition 8.1.4 together with associativity and skew-commutativity of the scalar product, which in our case is given by the Poincaré duality pairing (8.1). Such algebras have many beautiful and striking properties which are explained in the recent papers by Dubrovin [17] and Kontsevich and Manin [35]. In particular, they give rise to flat connections and integrable Hamiltonian systems. We shall discuss some of these structures in Section 8.5.

Example 8.1.6 (Complex projective space) Consider $\mathbb{C}P^n$ with its standard complex structure and the Fubini-Study metric with corresponding Kähler form ω. Let $L \in H_2(\mathbb{C}P^n)$ be the standard generator, represented by the line $\mathbb{C}P^1$. The first Chern class of $\mathbb{C}P^n$ is given by

$$c_1(L) = n + 1.$$

Hence, for reasons of dimension, the invariant $\Phi_{mL}(\alpha, \beta, \gamma)$ can only be non-zero when $m = 0, 1$. By Proposition 8.1.4 the case $m = 0$ corresponds to constant curves and gives the usual cup-product.

The minimal Chern number in this case is $N = n + 1$ and so the quantum cohomology groups are given by $QH^k(M) \cong \mathbb{Z}$ when k is even and $H^k(M) = \{0\}$ when k is odd. Let $a \in H^\ell(M)$ and $b \in H^m(M)$. If $\ell + m \leq 2n$ then the quantum cup product agrees with the ordinary cup product $a * b = a \cup b$. So the first interesting case is $\ell + m = 2n$. Consider the case where $a = p \in H^2(M)$ is the standard generator, defined by $p(L) = 1$, and $b = p^n \in H^{2n}(\mathbb{C}P^n)$. We shall prove that the quantum product $p * p^n$ is the generator $q \in QH^{2n+2}(\mathbb{C}P^n)$. In the notation of Remark 8.1.2 we must prove that $(p * p^n)_L = \mathbb{1} \in H^0(\mathbb{C}P^1)$. This means that $\int_{\mathrm{pt}}(p * p^n)_L = 1$. In view of (8.4) this integral is indeed given by

$$\int_{\mathrm{pt}} (p * p^n)_L = \Phi_L([\mathbb{C}P^{n-1}], \mathrm{pt}, \mathrm{pt}) = 1$$

where $[\mathbb{C}P^{n-1}] = \mathrm{PD}(p)$ and $\mathrm{pt} = \mathrm{PD}(p^n)$. Thus we have proved

$$p * p^n = q \in QH^{2n+2}(M).$$

The other non-zero terms $\Phi_L(\alpha, \beta, \gamma)$ correspond to relations of the form $p^k * p^\ell = p^{k+\ell-n-1}q$ for $k + \ell \geq n + 1$. It follows easily that quantum multiplication is associative in this case. In explicit terms the quantum cohomology ring $QH^*(M)$ has one generator $a_k \in QH^k(M)$ for every even integer k and the quantum deformation of the cup-product is now given by

$$a_k * a_\ell = a_{k+\ell}$$

for all $k, \ell \in 2\mathbb{Z}$. This can be expressed in the form

$$QH^*(\mathbb{C}P^n) = \frac{\mathbb{Z}[p, q, q^{-1}]}{\langle p^{n+1} = q, \; q^{-1}q = 1 \rangle}.$$

If we tensor with the polynomial ring $\mathbb{Z}[q]$ then the (non-periodic) quantum cohomology groups are given by

$$\widetilde{QH}^*(\mathbb{C}P^n) = \frac{\mathbb{Z}[p,q]}{\langle p^{n+1} = q \rangle}.$$

Specializing to $q = 0$, we recover the ordinary cohomology ring of $\mathbb{C}P^n$. $\square$

8.2   Associativity and composition rules

Our goal in this section is to give an outline of the proof of the following theorem. This result was first proved by Ruan and Tian [67, 68] and subsequently by Liu [38].

Theorem 8.2.1 (Ruan-Tian) *Let (M, ω) be a compact symplectic manifold which is monotone with minimal Chern number $N \geq 2$. Then the deformed cup product $*$ on the quantum cohomology group $QH^*(M)$ is associative.*

To prove this it suffices to show that the triple product

$$QH^j(M) \otimes QH^k(M) \otimes QH^\ell(M) \to QH^{j+k+\ell}(M)$$

which sends (a, b, c) to $(a * b) * c$ is skew symmetric with the usual sign conventions. Now there is an obvious such skew symmetric triple product defined as follows. Let $\alpha \in H_{2n-j}(M)$, $\beta \in H_{2n-k}(M)$, and $\gamma \in H_{2n-\ell}(M)$ be the Poincaré duals of $a \in H^j(M)$, $b \in H^k(M)$, $c \in H^\ell(M)$. Then define

$$a \star b \star c = \sum_A (a \star b \star c)_A q^{c_1(A)/N} \in QH^{j+k+\ell}(M)$$

by the formula

$$\int_\delta (a \star b \star c)_A = \Psi_A(\alpha, \beta, \gamma, \delta) \tag{8.5}$$

for $\delta \in H_{j+k+\ell-2Nd}(M)$. This is well defined because M satisfies hypotheses $(H4)$ and $(H5)$ in Section 7.4, and the classes α, β, γ, δ satisfy the dimension condition

$$\deg(\alpha) + \deg(\beta) + \deg(\gamma) + \deg(\delta) = 6n - 2c_1(A) \tag{8.6}$$

which is equivalent to (7.3).[2] The triple product (8.5) is obviously skew-symmetric and so Theorem 8.2.1 is a consequence of the following

Proposition 8.2.2 *If (M, ω) is monotone with minimal Chern number $N \geq 2$ then*

$$(a * b) * c = a \star b \star c$$

for all $a, b, c \in QH^*(M)$.

---

[2]Note that, already for dimensional reasons, we cannot use Φ_A in the definition (8.5) of our triple product, since this would give a product $QH^j \otimes QH^k \otimes QH^\ell \to QH^{j+k+\ell-2}$.

To understand the product $(a * b) * c$ we shall first examine the Poincaré dual of the cohomology class $a * b \in QH^{j+k}(M)$. This quantum cohomology class can be thought of as a sum

$$\sum_A (a * b)_A q^{c_1(A)/N}$$

where the cohomology class $(a * b)_A \in H^{j+k-2c_1(A)}(M)$ is defined by (8.4). We denote the Poincaré dual of $(a * b)_A$ by

$$\xi_A = \mathrm{PD}((a * b)_A) \in H_{2n+2c_1(A)-j-k}(M).$$

In view of (8.4) the homology class ξ_A is defined by the intersection property

$$\xi_A \cdot \gamma = \Phi_A(\alpha, \beta, \gamma)$$

where $\alpha = \mathrm{PD}(a) \in H_{2n-j}(M)$, $\beta = \mathrm{PD}(b) \in H_{2n-k}(M)$, and $\gamma \in H_{j+k-2c_1(A)}(M)$. With this notation the cohomology class

$$(a * b) * c = \sum_A ((a * b) * c)_A q^{c_1(A)/N}$$

with $((a * b) * c)_A \in H^{j+k+\ell-2c_1(A)}(M)$ is given by the formula

$$\int_\delta ((a * b) * c)_A = \sum_B \Phi_B(\xi_{A-B}, \gamma, \delta), \tag{8.7}$$

for $\delta \in H_{j+k+\ell-2c_1(A)}(M)$.

Our goal is to give a more direct description of the invariant $\Phi_B(\xi_A, \gamma, \delta)$. For this it is necessary to find a pseudo-cycle which represents the homology class ξ_A. Such a pseudo-cycle is in fact given by an evaluation map on a suitable moduli space of J-holomorphic curves. Roughly speaking, we may think of ξ_A as the union of all A-curves which go through α and β. More precisely, if z_1, z_2, z_3 are three distinct points in $\mathbb{C}P^1$ then ξ_A can represented by the pseudo-cycle $f_A : V \to M$ where

$$V = \{u \in \mathcal{M}(A, J) \,|\, u(z_1) \in \alpha, \, u(z_2) \in \beta\}, \qquad f_A(u) = u(z_3).$$

Observe that the manifold V need not be compact. However, it does carry a fundamental cycle because its boundary has codimension at least 2. The details of this are precisely as in Chapter 6. In fact, we may consider the evaluation map

$$e_{\mathbf{z}} = e_{A,J,\mathbf{z}} : \mathcal{M}(A, J) \to M^3$$

and choose representatives of α and β such that $\alpha \times \beta \times M$ is transverse to $e_{\mathbf{z}}$ and to all the evaluation maps appearing in the closure of $e_{\mathbf{z}}(\mathcal{M}(A, J))$. Then the manifold V is the preimage

$$V = e_{\mathbf{z}}^{-1}(\alpha \times \beta \times M)$$

and the evaluation map $f_A : V \to M$ defines a pseudo-cycle.

Now the invariant $\Phi_B(\xi_A, \gamma, \delta)$ counts the B-curves which go through ξ_A, γ, and δ. But every B-curve which meets ξ_A intersects some A-curve through α and β, and therefore determines a cusp-curve of type (A, B). Hence the invariant $\Phi_B(\xi_A, \gamma, \delta)$ counts the number of cusp curves of type (A, B) such that the A-component of meets α and β, and the B-component meets γ and δ. Here it is possible for A or B to be zero. For example if $A = 0$ then the B-curve meets $\alpha \cap \beta$, γ, and δ.

To formulate the invariant $\Phi_B(\xi_A, \gamma, \delta)$ more precisely, and to recover the symmetry between A and B, it is convenient to define an analogue of Ψ for pairs A, B of homology classes. Fix a point $z_0 \in \mathbb{C}P^1$ and consider the space of cusp-curves

$$\mathcal{M}(A, B, J) = \{(u, v) \in \mathcal{M}(A, J) \times \mathcal{M}(B, J) \mid u(z_0) = v(z_0)\}.$$

We have shown in Chapter 6 that this is a manifold of dimension $2n + 2c_1(A + B)$ for generic J (see Example 6.4.2). Fix a quadruple

$$\mathbf{z} = (z_1, z_2, z_3, z_4)$$

of distinct points in $\mathbb{C}P^1$ and define the evaluation map

$$e_{\mathbf{z}} = e_{A,B,J,\mathbf{z}} : \mathcal{M}(A, B, J) \to M^4$$

by the formula

$$e_{\mathbf{z}}(u, v) = (u(z_1), u(z_2), v(z_3), v(z_4)).$$

The following proposition asserts that for a generic almost complex structure J this map determines a pseudo-cycle. The proof is precisely the same as that of Theorem 5.4.1 and is left to the reader.

Proposition 8.2.3 *Assume (M, ω) is monotone with minimal Chern number $N \geq 2$ Fix $A, B \in \Gamma \subset H_2(M)$ and a quadruple $\mathbf{z} = (z_1, z_2, z_3, z_4)$ of distinct points in $\mathbb{C}P^1$. Then there exists a set $\mathcal{J}_{\mathrm{reg}} = \mathcal{J}_{\mathrm{reg}}(M, \omega, A, B, \mathbf{z}) \subset \mathcal{J}(M, \omega)$ of the second category such that the evaluation map $e_{\mathbf{z}} = e_{A,B,J,\mathbf{z}} : \mathcal{M}(A, B, J) \to M^4$ is a pseudo-cycle in the sense of Section 7.1 for every $J \in \mathcal{J}_{\mathrm{reg}}$. Moreover, the bordism class of $e_{A,B,J,\mathbf{z}}$ is independent of the choice of J.*

Given homology classes $\alpha, \beta, \gamma, \delta \in H_*(M)$ which satisfy the condition (8.6) choose a pseudo-cycle $f : V \to M^4$ which represents the class $\alpha \times \beta \times \gamma \times \delta$, and is transverse to the evaluation map $e_{A,B,J,\mathbf{z}}$ and to all the evaluation maps which occur in the description of the boundary of the set $e_{\mathbf{z}}(\mathcal{M}(A, B, J))$. Define the invariant $\Psi_{A,B}$ as the intersection number

$$\Psi_{A,B}(\alpha, \beta; \gamma, \delta) = e_{A,B,J,\mathbf{z}} \cdot f.$$

By Lemma 7.1.4 and Proposition 8.2.3 this integer is independent of the choice of the pseudo-cycle f and the almost complex structure J used to define it. In fact, the above discussion shows that

$$\Psi_{A,B}(\alpha, \beta; \gamma, \delta) = \Phi_B(\xi_A, \gamma, \delta)$$

where $\xi_A = \mathrm{PD}((a * b)_A)$. This formula can be used to express the invariant $\Psi_{A,B}$ entirely in terms of the original invariant Φ_A. To see this choose a basis $\{\varepsilon_i\}_i$ of the homology $H_*(M)$ and let $\{\phi_j\}_j$ be the dual basis with respect to the intersection pairing in the sense that

$$\phi_j \cdot \varepsilon_i = \delta_{ij}.$$

(Note that in the odd case care must be taken with the signs.)

Lemma 8.2.4

$$\Psi_{A,B}(\alpha, \beta; \gamma, \delta) = \sum_i \Phi_A(\alpha, \beta, \varepsilon_i)\Phi_B(\phi_i, \gamma, \delta).$$

Proof: Denote $e_i = \mathrm{PD}(\varepsilon_i)$, $f_j = \mathrm{PD}(\phi_j)$, $a = \mathrm{PD}(\alpha)$, etc. Then $\langle f_j, e_i \rangle = \delta_{ij}$ and we obtain

$$(a * b)_A = \sum_i \langle (a * b)_A, e_i \rangle f_i = \sum_i \Phi_A(\alpha, \beta, \varepsilon_i) f_i.$$

Similarly,

$$(c * d)_B = \sum_i \langle f_i, (c * d)_B \rangle e_i = \sum_i \Phi_A(\phi_i, \gamma, \delta) e_i$$

Hence

$$\langle (a * b)_A, (c * d)_B \rangle = \sum_i \Phi_A(\alpha, \beta, \varepsilon_i)\Phi_B(\phi_i, \gamma, \delta).$$

With $\xi_A = \mathrm{PD}((a * b)_A)$ as above the left hand side is given by

$$\langle (a * b)_A, (c * d)_B \rangle = \Phi_B(\xi_A, \gamma, \delta) = \Psi_{A,B}(\alpha, \beta; \gamma, \delta).$$

This proves the lemma. $\qquad\qquad\square$

Now we may write the formula (8.7) for $(a * b) * c$ in the form

$$\int_\delta ((a * b) * c)_A = \sum_B \Psi_{B, A-B}(\alpha, \beta; \gamma, \delta), \qquad (8.8)$$

for $\delta \in H_{j+k+m-2c_1(A)}(M)$. Note that the classes α, β, γ, δ satisfy the dimension condition (8.6). Comparing the formulae (8.8) and (8.5), we see that the proof of Proposition 8.2.2, and hence that of Theorem 8.2.1, reduces to the following identity.

Lemma 8.2.5 *Assume (M, ω) is monotone with minimal Chern number $N \geq 2$ and let $A \in H_2(M)$. Then*

$$\Psi_A(\alpha_1, \alpha_2, \alpha_3, \alpha_4) = \sum_B \Psi_{B, A-B}(\alpha_1, \alpha_2; \alpha_3, \alpha_4)$$

for cohomology classes $\alpha_j \in H_*(M)$ with $\sum_j \deg(\alpha_j) = 6n - 2c_1(A)$.

Combining Lemma 8.2.5 and Lemma 8.2.4 we obtain the following composition rule which was stated by Ruan and Tian [67] in much more generality.

Corollary 8.2.6

$$\Psi_A(\alpha_1, \alpha_2, \alpha_3, \alpha_4) = \sum_B \sum_i \Psi_{A-B}(\alpha_1, \alpha_2, \varepsilon_i)\Psi_B(\phi_i, \alpha_3, \alpha_4).$$

To understand why Lemma 8.2.5 might be true consider the evaluation map

$$e_{\mathbf{z}} : \mathcal{M}(A, J) \to M^4$$

for the quadruple $\mathbf{z} = (0, 1, \infty, z) \in (\mathbb{C}P^1)^4$ where the point z varies in $\mathbb{C} - \{0, 1\}$. The elements $w \in \mathcal{M}(A, J)$ with $e_{\mathbf{z}}(w) \in f$ are A-curves which satisfy

$$w(\infty) \in \alpha_1, \quad w(1) \in \alpha_2, \quad w(0) \in \alpha_3, \quad w(z) \in \alpha_4. \tag{8.9}$$

Here we assume that the product cycle $\alpha_1 \times \alpha_2 \times \alpha_3 \times \alpha_4$ is transverse to the evaluation map $e_{\mathbf{z}}$ (see Remark 7.2.1). The invariant $\Psi_A(\alpha_1, \alpha_2, \alpha_3, \alpha_4)$ counts the number of curves which satisfy this condition. The goal is to show that when z is sufficiently close to 0 there is a one-to-one correspondence between such A-curves and the connected pairs of curves which are counted by $\Psi_{A-B,B}(\alpha_1, \alpha_2; \alpha_3, \alpha_4)$.

Consider what happens for a sequence $w_\nu \in \mathcal{M}(A, J)$ which satisfies (8.9) when the corresponding point z_ν tends to 0. Then either the sequences $w_\nu(z_\nu)$ and $w_\nu(0)$ converge to the same point which therefore must lie in $\alpha_3 \cap \alpha_4$, or they converge to different points. In the former case the limit is an A-curve which contributes to

$$\Phi_A(\alpha_1, \alpha_2, \alpha_3 \cap \alpha_4) = \Psi_{A,0}(\alpha_1, \alpha_2; \alpha_3, \alpha_4).$$

In the latter case, the two nearby points 0 and z get mapped to widely separated points, and so the derivative of w_ν must blow up somewhere near 0. When one rescales at the blow-up point as in the proof of Theorem 4.3.2, one will obtain a bubble which meets α_3 and α_4 and lies in some class B. Generically, in view of Lemma 6.5.1, there will only be this one bubble. Hence the restriction of w_ν to compact subsets of $\mathbb{C}P^1 - \{0\}$ will converge to a curve $w_\infty = u \in \mathcal{M}(A - B, J)$ which still takes ∞ to α_1 and 1 to α_2. Thus the limiting cusp-curve contributes to

$$\Psi_{A-B,B}(\alpha_1, \alpha_2; \alpha_3, \alpha_4).$$

Included here is the possibility that the limit curve u is constant, which corresponds to the term

$$\Phi_A(\alpha_1 \cap \alpha_2, \alpha_3, \alpha_4) = \Psi_{0,A}(\alpha_1, \alpha_2; \alpha_3, \alpha_4).$$

This argument shows that if z is sufficiently close to 0 every element w which is counted in the computation of the invariant $\Psi_A(\alpha_1, \alpha_2, \alpha_3, \alpha_4)$ has a counterpart which contributes to the sum

$$\sum_B \Psi_{A-B,B}(\alpha_1, \alpha_2; \alpha_3, \alpha_4).$$

To complete the proof one must show that, conversely, every pair of intersecting curves $(u, v) \in \mathcal{M}(A - B, B, J)$ with $u(\mathbb{C}P^1)$ intersecting α_1 and α_2 and $v(\mathbb{C}P^1)$ intersecting α_3 and α_4 determines a unique curve $w \in \mathcal{M}(A, J)$ which satisfies (8.9) provided that z is sufficiently close to 0. A proof using Floer's original technique involving infinite cylindrical ends has been worked out by Liu [38]. We shall present a more direct argument in Appendix A. Ruan and Tian take a somewhat different approach to this problem which involves the perturbed Cauchy-Riemann equations. Their argument is outlined in [67] and the details are carried out in [68].

Remark 8.2.7 An observant reader will notice that no mention has been made above of the effect of multiply-covered curves. It would be possible for the limiting curve to be a cusp-curve with one or both components multiply-covered. However, the fact that each component only contains two marked points means that such an event occurs with high codimension and so may be ignored. This may be proved by the methods of Chapter 6. $\qquad\square$

Example 8.2.8 We have already seen that in $\mathbb{C}P^2$,

$$\Psi_{2L}(\mathrm{pt},\mathrm{pt},\mathrm{pt},\mathrm{pt}) = 1.$$

Since $\alpha_i \cap \alpha_j = \emptyset$ here, there is no contribution from decompositions $2L = A + B$ where one of the classes A or B is zero. According to Lemma 8.2.5 we should therefore have

$$\Psi_{L,L}(\mathrm{pt},\mathrm{pt};\mathrm{pt},\mathrm{pt}) = 1.$$

Here the left hand side is the number of pairs of intersecting lines L_1, L_2 with the first two points on L_1 and the second two on L_2. Clearly, there is a unique such pair. In a similar way one can check this lemma for cases such as $\Psi_{2L}(\mathrm{pt},\mathrm{pt},\mathrm{pt},\mathrm{line})$ in $\mathbb{C}P^3$ or $\Psi_L(\mathrm{pt},\mathrm{line},\mathrm{line},\mathrm{line})$ in $\mathbb{C}P^2$. It is also easy to check that Lemma 8.2.4 holds in these cases. $\qquad\square$

8.3 Flag manifolds

In a beautiful recent paper [24], Givental and Kim have computed the quantum cohomology ring of the flag manifold F_{n+1}, and related it to the Toda lattice. To do this they made certain assumptions about the properties of quantum cohomology which so far have not been established. However, their formula has been verified by Ciocan-Fontanine in a recent paper [10]. In [3] Astashkevich and Sadov generalize these results (in a heuristic way) to partial flag manifolds. The extreme case of Grassmannians $G(k,n)$ will be discussed in the next section.

Recall that the flag manifold F_{n+1} is the space of all sequences of subspaces

$$E_1 \subset E_2 \subset \cdots \subset E_n$$

of $\mathbb{C}^{n+1}$ with $\dim_{\mathbb{C}} E_j = j$. This manifold is simply connected and carries a natural complex structure. The minimal Chern number is 2. Its cohomology ring is generated by the first Chern classes $u_j \in H^2(M,\mathbb{Z})$ of the canonical line bundles

$$L_j \to F_{n+1}$$

with fiber E_{j+1}/E_j for $j = 0, \dots, n$. Since the sum $L_0 \oplus \cdots \oplus L_n$ is trivial, these classes are related by the condition

$$u_0 + \cdots + u_n = 0.$$

The full cohomology ring of F_{n+1} is the quotient

$$H^*(F_{n+1}) = \frac{\mathbb{Z}[u_0,\dots,u_n]}{\langle \sigma_1(u),\dots,\sigma_{n+1}(u)\rangle}$$

where the $\sigma_j(u)$ denote the elementary symmetric functions. To understand this, observe first that the cohomology classes $c_j = \sigma_j(u) \in H^{2j}(F_{n+1})$ are the Chern classes of the trivial bundle $L_0 \oplus \cdots \oplus L_n$ and so must obviously be zero. The above formula asserts that these obvious relations are the only ones and that all the cohomology of F_{n+1} is generated by the classes u_j via the cup product.

The cohomology ring can also be expressed in terms of the basis

$$p_j = u_j + \cdots + u_n$$

with $j = 1, \ldots, n$. In this basis the first Chern class of the tangent bundle TF_{n+1} is given by

$$c_1 = 2(p_1 + \cdots + p_n).$$

The classes p_j can be represented by forms which are Kähler with respect to the obvious complex structure J on F_{n+1}. In fact a cohomology class $a = \sum_j \lambda_j p_j$ can be represented by a Kähler form precisely if the coefficients λ_j are all nonnegative and their sum is positive. In particular, there exists a Kähler form with respect to which the manifold F_{n+1} is monotone and hence quantum cohomology is well defined. It also follows that $p_j(A) > 0$ for every J-effective homology class A and every j, since J is tamed by the Kähler form in class p_j.

Define the degrees d_j of a homology class A in terms of the homomorphisms $p_j : H_2(M, \mathbb{Z}) \to \mathbb{Z}$. In other words, a homology class $A \in H_2(M)$ is uniquely determined by the integers

$$d_j = p_j(A)$$

for $j = 1, \ldots, n$ and, conversely, any such set of integers determines a homology class $A = A_d$. Above we associated to a class $A \in H_2(M)$ the monomial q^d where $c_1(A) = Nd$, N is the minimal Chern number, and $\deg q = 2N$. Now, in order to capture the full structure of $H_2(M)$, we take n auxiliary variables $q_1, \ldots, q_n$ each of degree $2N = 4$, which represent the basis of $H_2(F_{n+1})$ dual to the basis $p_1, \ldots, p_n$ of $H^2(F_{n+1})$. We then represent the class A by the the monomial

$$q^d = q_1^{d_1} \cdots q_n^{d_n}$$

where $d = (d_1, \ldots, d_n) \in \mathbb{Z}^n$ and $d_j = p_j(A)$. Since $c_1 = 2(p_1 + \ldots + p_n)$, this is consistent with our previous definitions when we identify all q_j with q.

Since $d_j = p_j(A) > 0$ for every effective homology class A it follows that the pairing

$$\langle a * b, \gamma \rangle = \sum_d \Phi_{A_d}(\alpha, \beta, \gamma) q^d$$

(with $A_d \in H_2(M)$ determined by $p_j(A_d) = d_j$ for $d \in \mathbb{Z}^n$) is a polynomial of degree $\ell + k - j$ for each $a \in H^k(F_{n+1})$, $b \in H^\ell(F_{n+1})$, and $\gamma \in H_j(F_{n+1})$. Hence, we may take the coefficient ring Λ for quantum cohomology to be the polynomial ring $\mathbb{Z}[q_1, \ldots, q_n]$. In other words, we define

$$\widetilde{QH}^*(F_{n+1}) = H^*(F_{n+1}) \otimes \Lambda = H^*(F_{n+1}) \otimes \mathbb{Z}[q_1, \ldots, q_n].$$

Since the u_i are multiplicative generators of $H^*(F_{n+1})$, the quantum cohomology ring $\widetilde{QH}^*(F_{n+1})$ is isomorphic to some quotient of $\mathbb{Z}[u_0, \ldots, u_n, q_0, \ldots, q_n]$.

In [24] Givental and Kim conjectured[3] that there is a natural isomorphism

$$\widetilde{QH}^*(F_{n+1}) \cong \frac{\mathbb{Z}[u_0,\ldots,u_n,q_0,\ldots,q_n]}{\mathcal{I}}$$

where $\mathcal{I} \subset \mathbb{Z}[u_0,\ldots,u_n,q_0,\ldots,q_n]$ denotes the ideal generated by the coefficients of the characteristic polynomial of the matrix

$$A_n = \begin{pmatrix} u_0 & q_1 & 0 & 0 & \cdots & \cdots & 0 \\ -1 & u_1 & q_2 & 0 & & & \vdots \\ 0 & -1 & u_2 & q_3 & \ddots & & \vdots \\ 0 & 0 & -1 & \ddots & \ddots & \ddots & \vdots \\ \vdots & & \ddots & \ddots & \ddots & \ddots & 0 \\ \vdots & & & \ddots & \ddots & u_{n-1} & q_n \\ 0 & \cdots & \cdots & \cdots & 0 & -1 & u_n \end{pmatrix}.$$

More explicitly consider the polynomials

$$c_j = \Sigma_j(u,q)$$

determined by the formula

$$\det(A_n + \lambda) = \lambda^{n+1} + c_1\lambda^n + \cdots + c_n\lambda + c_{n+1}.$$

The ideal $\mathcal{I}$ is generated by these functions Σ_j. In the case $q = 0$ these are the elementary symmetric functions $\sigma_j(u) = \Sigma_j(u,0)$ and thus the ordinary cohomology ring appears as expected when we specialize to $q = 0$. In other words the classical Chern classes are given by the elementary symmetric functions and the polynomials Σ_j can be regarded as the **quantum deformations of the Chern classes**. A similar phenomenon appeared in the discussion of Grassmannians in [79].

This result has some very interesting connections with completely integrable Hamiltonian systems which we now explain. We will see that the quantum Chern classes Σ_j are the Poisson commuting integrals of the Toda lattice (cf. [56]).

Recall that the Toda lattice is a Hamiltonian differential equation for $n + 1$ unit masses on the real line in the positions $x_0, x_1, \ldots, x_n$. If

$$u_j = p_j - p_{j+1} = y_j = \dot{x}_j$$

is the momentum of the particle at position x_j and if

$$q_j = e^{x_j - x_{j-1}},$$

then the Hamiltonian function of the Toda lattice is given by

$$H(x,y) = \tfrac{1}{2}\mathrm{trace}(A_n^2) = \tfrac{1}{2}\sum_{j=0}^{n} y_j^2 - \sum_{j=1}^{n} e^{x_j - x_{j-1}}.$$

---

[3]The computation of Givental and Kim assumes an equivariant version of quantum cohomology which so far has not been established. The argument of Ciocan-Fontanine avoids this difficulty.

(The reversal of the usual signs corresponds to considering forces which repel when $x_0 < x_1 < \cdots < x_n$.) In [56] Moser discovered that the functions $F_j(x,y) = \operatorname{trace}(A_n^j)$ for $j = 1, \ldots, n+1$ form a complete set of Poisson commuting integrals for this system, where the matrix A_n is as defined above. As a result the ideal $\mathcal{I}$ generated by the quantum Chern classes Σ_j is invariant under Poisson brackets.

To state this more precisely, observe that the standard symplectic structure $\omega = \sum_{j=0}^{n} dx_j \wedge dy_j$, when restricted to the set $x_0 + \cdots + x_n = 0$ (zero center of mass) and $y_0 + \cdots + y_n = 0$ (zero momentum) and written in terms of the variables $q_1, \ldots, q_n, p_1, \ldots, p_n$, takes the form

$$\omega = \frac{dq_1}{q_1} \wedge dp_1 + \cdots + \frac{dq_n}{q_n} \wedge dp_n.$$

The Poisson structure is to be understood with respect to this symplectic form.

We can give the following geometric interpretation of ω. Consider the complex torus

$$\mathbb{T}_{\mathbb{C}} = H^2(F_{n+1}; \mathbb{C}/\mathbb{Z}) = H^2(F_{n+1}; \mathbb{C})/H^2(F_{n+1}; \mathbb{Z})$$

which is parametrized by the coordinate functions

$$q_j(a) = e^{2\pi i \langle a, A_j \rangle}.$$

The cotangent bundle of $\mathbb{T}_{\mathbb{C}}$ can be naturally identified with

$$T^* \mathbb{T}_{\mathbb{C}} = H^2(F_{n+1}, \mathbb{C}/\mathbb{Z}) \times H_2(F_{n+1}, \mathbb{C})$$

where the $p_j : H_2(M, \mathbb{C}) \to \mathbb{C}$ are to be understood as coordinate functions on the cotangent space. (These coordinate functions have a geometric meaning in terms of the cone of Kähler forms as explained above.) Then the form ω can be understood as a symplectic structure on the cotangent bundle $T^* \mathbb{T}_{\mathbb{C}}$. In this interpretation the above Hamiltonian system with particles $x_0, \ldots, x_n$, when restricted to the set of zero center of mass and zero momentum, lives on the imaginary part of $T^* \mathbb{T}$.

This leads to a geometric interpretation of the quantum cohomology itself. The ideal

$$\mathcal{J} \subset \mathbb{C}[p_1, \ldots, p_n, q_1, \ldots, q_n]$$

which corresponds to $\mathcal{I}$ determines an algebraic variety

$$\mathcal{L} \subset T^* \mathbb{T}_{\mathbb{C}}.$$

This variety is defined as the common zero set of the polynomials in $\mathcal{J}$. Hence the quantum cohomology ring

$$\widetilde{QH}^*(M, \mathbb{C}) = \frac{\mathbb{C}[p_1, \ldots, p_n, q_1, \ldots, q_n]}{\mathcal{J}}$$

with complex coefficients can be interpreted as the space of functions on $\mathcal{L}$. The invariance of the ideal $\mathcal{J}$ under Poisson brackets translates into the condition that the variety $\mathcal{L}$ is Lagrangian. It should therefore have a generating function and this, according to Givental and Kim, should have a bearing on the question of mirror symmetry.

8.4 Grassmannians

The quantum cohomology ring of the Grassmannian $G(k, N)$ has been studied by Bertram, Daskalopoulos, and Wentworth in [6], Witten in [87], and Siebert and Tian in [79]. In [87] Witten also relates the quantum cohomology of $G(k, N)$ to the Verlinde algebra of representations of $U(k)$. His explanation for this relation is based on a beautiful, but heuristic, consideration involving path integrals, and so far there is no rigorous proof.

Denote by $G(k, N)$ the Grassmannian of k-planes in $\mathbb{C}^N$. Thus a point in $G(k, N)$ is a k-dimensional subspace $V \subset \mathbb{C}^N$. A **unitary frame** of V is a matrix $B \subset \mathbb{C}^{N \times k}$ such that

$$V = \operatorname{im} B, \qquad B^* B = \mathbb{1}_{k \times k}.$$

Two such frames B and B' represent the same subspace V if there exists a unitary matrix $U \in U(k)$ such that $B' = BU$. Hence the Grassmannian can be identified with the quotient space

$$G(k, N) = P/U(k)$$

where $P \subset \mathbb{C}^{N \times k}$ denotes the set of unitary k-frames. A moment's thought shows that the Grassmannian has real dimension

$$\dim G(k, N) = 2k(N - k).$$

It can also be interpreted as a symplectic quotient. The space $\mathbb{C}^{N \times k}$ is a symplectic manifold and the natural action of the group $U(k)$ on this space is Hamiltonian with moment map $\mu(B) = B^* B/2i$. The Grassmannian now appears as the quotient $G(k, N) = \mu^{-1}(\mathbb{1}/2i)/U(k)$.

Now there are two natural complex vector bundles, $E \to G(k, N)$ of rank k and $F \to G(k, N)$ of rank $N - k$, whose Whitney sum is naturally isomorphic to the trivial bundle $G(k, N) \times \mathbb{C}^N$. The fiber of E at the point $V \in G(k, N)$ is just the space V itself and the fiber of F is the quotient $\mathbb{C}^N/V$:

$$E_V = V, \qquad F_V = \mathbb{C}^N/V.$$

Denote the Chern classes of the dual bundles E^* and F^* by

$$x_j = c_j(E^*) \in H^{2j}(G(k, N)), \qquad y_j = c_j(F^*) \in H^{2j}(G(k, N)).$$

These classes generate the cohomology of $G(k, N)$. Since $E \oplus F$ is isomorphic to the trivial bundle there are obvious relations

$$\sum_{i=0}^{j} x_i y_{j-i} = 0$$

for $j = 1, \ldots, N$. For $j > N$ this equation is trivially satisfied. For $j = 1, \ldots, N - k$ it determines the classes y_j inductively as functions of $x_1, \ldots, x_k$ via

$$y_j = -x_1 y_{j-1} - \cdots - x_{j-1} y_1 - x_j, \qquad j = 1, \ldots, N - k.$$

For $j > N - k$ the classes y_j vanish and this determines relations of the x_j. These are the only relations and hence the cohomology ring of the Grassmannian can be identified with the quotient

$$H^*(\mathrm{G}(k,N),\mathbb{C}) \cong \frac{\mathbb{C}[x_1,\ldots,x_k]}{\langle y_{N-k+1},\ldots,y_N\rangle}.$$

Moreover, the first Chern class of the tangent bundle is given by $c_1(T\mathrm{G}(k,N)) = Nx_1$ and so the minimal Chern number is N. The following theorem was essentially proved by Witten [87]. Independently, a rigorous proof with all details was worked out by Siebert and Tian [79].

Theorem 8.4.1 (Siebert-Tian, Witten) *The quantum cohomology of the Grassmannian is isomorphic to the ring*

$$QH^*(\mathrm{G}(k,N)) \cong \frac{\mathbb{C}[x_1,\ldots,x_k,q]}{\langle y_{N-k+1},\ldots,y_{N-1},y_N+(-1)^{N-k}q\rangle}.$$

Here x_j is a generator of degree $2j$ and q is a generator of degree $2N$. The relation $y_N + (-1)^{N-k}q = 0$ can also be written in the form

$$x_k y_{N-k} = (-1)^{N-k}q.$$

Proof: We remark first that, by an easy induction argument, the classes $x_1,\ldots,x_k$ still generate the quantum cohomology of $\mathrm{G}(k,N)$. This means that every cohomology class can be expressed as a linear combination of quantum products of the x_i. To prove this one uses induction over the degree and the fact that the difference $x*y - x\cup y$ is a sum of terms of lower degree that $x\cup y$. (See Lemma 2.1 in [79] for details.) Now this same argument shows that the original relations in the classical cohomology ring become relations in quantum cohomology by adding certain lower order terms. It follows again by induction over the degree that these new relations generate the ideal of relations in quantum cohomology. (See Theorem 2.2 in [79] for details.)

In view of these general remarks we must compute the quantum deformations of the defining relations in the cohomology ring. In this we follow Witten's argument in [87]. The quantum cup product of the classes x_i and y_j is a power series of the form

$$x_i * y_j = \sum_d (x_i * y_j)_d q^d$$

where $(x_i * y_j)_d \in H^{2i+2j-2Nd}(\mathrm{G}(k,N))$ and $(x_j * y_j)_0 = x_i \cup y_j$. It follows that the quantum cup-product $x_i * y_j$ must agree with the ordinary cupproduct $x_i \cup y_j$ unless $i = k$ and $j = N - k$. Now the relations $y_j = 0$ for $j = N-k+1,\ldots,N-1$ only involve the products $x_i y_j$ with either $i < k$ or $j < N - k$ and hence they remain valid in quantum cohomology. However the relation $y_N = 0$ involves the product $x_k y_{N-k}$ and the only nontrivial contribution to the quantum deformation of this product is the term $(x_k * y_{N-k})_1 \in H^0(\mathrm{G}(k,N))$. We claim that

$$(x_k * y_{N-k})_1 = (-1)^{N-k}\mathbb{1}. \tag{8.10}$$

To see this we must examine the moduli space $\mathcal{M}(L,i)$ of holomorphic curves $u : \mathbb{C}P^1 \to \mathrm{G}(k,N)$ of degree $\deg(u) = c_1(L) - 1$. The space of such curves has formal dimension

$$\dim \mathcal{M}(L,i) = 2k(N-k) + 2N.$$

Every holomorphic curve $u \in \mathcal{M}(L, i)$ is of the form

$$u([z_0 : z_1]) = \mathrm{span}\,\{z_0 v_0 + z_1 v_1, v_2, \ldots, v_k\} \tag{8.11}$$

where the vectors $v_0, \ldots, v_k \in \mathbb{C}^N$ are linearly independent. Note that these maps form indeed a space of dimension $2N(k+1) - 2k^2$ which is in accordance with the dimension formula for $\mathcal{M}(L, i)$. Moreover, one can check that all these curves are regular. We must prove that

$$\Phi_L(\xi_k, \eta_{N-k}, \mathrm{pt}) = (-1)^{N-k}$$

where $\xi_k = \mathrm{PD}(x_k)$ and $\eta_{N-k} = \mathrm{PD}(y_{N-k})$. Now the Poincaré dual of the top-dimensional Chern class of a vector bundle can be represented by the zero set of a generic section. For example fix a vector $w \in \mathbb{C}^n$ and consider the section $\mathrm{G}(k, N) \to E$ which assigns to every k-plane $V \subset \mathbb{C}^N$ the restriction of the functional $v \mapsto w^* v$ to V. This section is transverse to the zero section and its zero set is the submanifold

$$X = \{V \in \mathrm{G}(k, N) \,|\, w^* v = 0 \,\forall\, v \in V\}.$$

This submanifold is a copy of $\mathrm{G}(k, N-1)$ in $\mathrm{G}(k, N)$ and represents the class $\xi_k \in H^{2k}(\mathrm{G}(k, N))$. Now fix a vector $v_0 \in \mathbb{C}^N$ and consider the section $\mathrm{G}(k, N) \to F$ which assigns to every $V \in \mathrm{G}(k, N)$ the equivalence class $[v_0] \in \mathbb{C}^N/V = F_V$. The zero set of this section is the submanifold

$$Y = \{V \in \mathrm{G}(k, N) \,|\, v_0 \in V\}.$$

This submanifold with its natural orientation represents the Poincaré dual of the top Chern class $c_{N-k}(F) = (-1)^{N-k} c_{N-k}(F^*)$. In summary,

$$[X] = \xi_k, \qquad [Y] = (-1)^{N-k} \eta_{N-k}.$$

Finally fix any point $V_0 \in \mathrm{G}(k, N) - X$ such that $v_0 \notin V_0$ and choose a basis $v_1, \ldots, v_k$ of V_0 such that the vectors $v_2, \ldots, v_k$ and $v_0 + v_1$ are perpendicular to w. Then the curve $u([z_0 : z_1]) = \mathrm{span}\{z_0 v_0 + z_1 v_1, v_2, \ldots, v_k\}$ satisfies

$$u([0 : 1]) = V_0, \qquad u([1 : 1]) \in X, \qquad u([1 : 0]) \in Y.$$

Moreover, the intersection number at u is 1 and it is easy to see that there is no other curve of degree 1 whose image intersects X, Y, and V_0. Hence

$$\Phi_L([X], [Y], \mathrm{pt}) = 1$$

and this proves the formula (8.10). It follows that in quantum cohomology the deformed relations are $y_j = 0$ for $N - k + 1 \leq j \leq N - 1$ and $y_N + (-1)^{N-k} q = 0$. This proves the theorem. $\qquad\square$

More details of the proof can be found in the paper [79] by Siebert and Tian. For example they give a proof of the fact that every holomorphic curve of degree 1 in $\mathrm{G}(k, N)$ must be of the form (8.11) and that they are regular in the sense that the associated Cauchy-Riemann operator D_u is surjective.

Theorem 8.4.1 shows once again how the quantum cohomology reduces to the classical cohomology ring if we consider q to be a complex number rather than a variable and specialize to $q = 0$. More explicitly, if we define the Chern polynomials

$$c_t(E^*) = \sum_{i=1}^{k} x_i t^i, \qquad c_t(F^*) = \sum_{j=1}^{N-k} y_j t^j$$

then the classical cohomology ring is determined by the relation

$$c_t(E^*)c_t(F^*) = 1$$

whereas the quantum cohomology ring is determined by

$$c_t(E^*)c_t(F^*) = 1 + (-1)^{N-k} q t^N \tag{8.12}$$

where $q \in \mathbb{C}$.

Landau-Ginzburg formulation

The relations in classical cohomology can be generated by the derivatives of a single function $W_0 = W_0(x_1, \ldots, x_k)$. The same is true for the relations in quantum cohomology and the corresponding function $W = W(x_1, \ldots, x_k)$ is called the **Landau-Ginzburg potential**. In our discussion of this approach we follow closely the exposition of Witten in [87].

Consider the polynomial

$$c_t(E^*) = \sum_{i=1}^{k} x_i t^i$$

where $x_i = c_i(E^*) \in H^{2i}(\mathrm{G}(k, N))$. Define polynomials $y_j = y_j(x_1, \ldots, x_k)$ for $j \geq 0$ by the formula

$$\frac{1}{c_t(E^*)} = \sum_{j \geq 0} y_j t^j.$$

Then the classical cohomology ring of the Grassmannian is described by the relations $y_j(x) = 0$ for $N - k + 1 \leq j \leq N$. These relations imply $y_j(x) = 0$ for $j > N$ and, of course, for $1 \leq j \leq N - k$ the classes y_j are the Chern classes of F^*. In the following we shall, however, not impose the condition $y_j(x) = 0$ and then the power series $\sum_j y_j t^j$ may no longer be a polynomial.

Consider the holomorphic functions $U_r = U_r(x_1, \ldots, x_k)$ defined by the equation

$$-\log c_t(E^*) = \sum_{r \geq 0} U_r(x) t^r.$$

Differentiating this expression with respect to x_j we see that

$$-\frac{t^j}{c_t(E^*)} = \sum_{r \geq 0} \frac{\partial U_r}{\partial x_j} t^r.$$

Comparing coefficients we find $\partial U_r / \partial x_j = -y_{r-j}$ for $1 \le j \le k$. In particular, when $r - j$ is negative this formula implies the vanishing of the corresponding derivative of U_r. The most interesting case is $r = N + 1$. With

$$W_0 = (-1)^{N+1} U_{N+1}$$

we obtain

$$\frac{\partial W_0}{\partial x_j} = (-1)^N y_{N+1-j}$$

for $1 \le j \le k$. Hence the defining relations for the classical cohomology of the Grassmannian can be written in the form

$$dW_0 = 0.$$

Now consider the function

$$W = W_0 + (-1)^k q x_1$$

where q is a fixed complex number. Then the condition $dW = 0$ is equivalent to

$$y_{N-k+1} = 0, \quad \cdots \quad , y_{N-1} = 0, \quad y_N + (-1)^{N-k} q = 0$$

and these are precisely the defining relations for the quantum cohomology ring of $G(k, N)$. Now the higher coefficients of the power series $\sum_j y_j t^j$ will no longer vanish. Comparing the coefficients up to order N in the equation $c_t(E^*) \cdot (\sum_{j \ge 0} y_j t^j) = 1$ we obtain

$$\left(\sum_{j=1}^{k} x_i t^i \right) \cdot \left(\sum_{j=1}^{N-k} y_j t^j \right) = 1 + (-1)^{N-k} q t^N$$

and this agrees with (8.12). The function W is called the **Landau-Ginzburg potential**. It can be conveniently expressed in terms of the roots $\lambda_1, \ldots, \lambda_k$ of the Chern polynomial

$$c_t(E^*) = \sum_{i=1}^{k} x_i t^i = \prod_{i=1}^{k} (1 + \lambda_i t), \tag{8.13}$$

namely

$$W(\lambda_1, \ldots, \lambda_k) = \sum_{i=1}^{k} \left(\frac{\lambda_i^{N+1}}{N+1} + (-1)^k q \lambda_i \right). \tag{8.14}$$

Geometrically, the quantum cohomology ring of $G(k, N)$ can be interpreted as the ring of polynomials in the variables $x_1, \ldots, x_k$ restricted to the zero set of dW. Now the function $W : \mathbb{C}^k \to \mathbb{C}$ has only finitely many critical points and the equivalence class of any polynomial $f \in \mathbb{C}[x_1, \ldots, x_k]$ with respect to the ideal

$$\mathcal{J}_q = \langle y_{N-k+1}, \ldots, y_{N-1}, y_N + (-1)^{N-k} q \rangle = \left\langle \frac{\partial W}{\partial x_1}, \ldots, \frac{\partial W}{\partial x_k} \right\rangle$$

is determined by the values of f at the critical points of W. This gives rise to localization formulae such as

$$I(f) = \frac{(-1)^{k(k-1)/2}}{k!} \sum_{dW(x)=0} \det \left(\frac{\partial^2 W}{\partial x_i \partial x_j} \right)^{-1} f(x) \tag{8.15}$$

for every polynomial f of degree $2k(N-k)$. Here the variable x_i is understood to be of degree $2i$ and the functional $I(f)$ denotes the integral of the differential form $\omega_f \in \Omega^{2k(N-k)}(\mathrm{G}(k,N))$ associated to f under the isomorphism $H^*(\mathrm{G}(k,N),\mathbb{C}) = \mathbb{C}[x_1,\dots,x_k]/\langle dW_0\rangle$.

Exercise 8.4.2 Prove that

$$\int_{\mathrm{G}(k,N)} c_k(E^*)^{N-k} = 1$$

by considering intersection points of $N-k$ copies of $\mathrm{G}(k,N-1)$ in $\mathrm{G}(k,N)$. Now check the formula (8.15) by applying it to the polynomial

$$f(x) = x_k{}^{N-k} = \prod_{i=1}^{k} \lambda_i{}^{N-k}$$

which represents the class $c_k(E^*)^{N-k}$. The general case follows from this because $H^{2k(N-k)}(\mathrm{G}(k,N))$ is one dimensional. **Hint:** First use change of variables and the residue calculus to express the sum (8.15) as a contour integral of the form

$$I(f) = \frac{(-1)^{k(k-1)/2}}{k!(2\pi i)^k} \int_{|\lambda_j|=R} \frac{f\cdot\prod_{i<j}(\lambda_i-\lambda_j)^2}{\prod_i \partial W/\partial\lambda_i}\, d\lambda_1\dots d\lambda_k$$

where R is large. Then note that the term $\partial W/\partial\lambda_i = \lambda_i{}^N + (-1)^k q$ in the denominator can be replaced by $\lambda_i{}^N - \lambda_i{}^{N-k}$ without changing the value of the integral. Now use the residue calculus again to evaluate the new integral. For details see [87] and [79]. $\qquad\qquad\qquad\qquad\qquad\qquad\qquad\qquad\qquad\qquad\qquad\qquad\qquad\Box$

Relation with the Verlinde algebra

There is a beautiful relation between the classical cohomology ring of the Grassmannian $\mathrm{G}(k,N)$ and the algebra of representations of the unitary group $\mathrm{U}(k)$ furnished by the Chern character. First note that the isomorphism classes of finite dimensional representations $\rho : \mathrm{U}(k) \to \mathrm{Aut}(V_\rho)$ form an algebra $\mathcal{R}_k$ with addition given by direct sum and multiplication by tensor product. More precisely one should form the free $\mathbb{Z}$-module generated by the isomorphism classes of irreducible representations. Now there is an algebra homomorphism

$$\mathcal{R}_k \to H^*(\mathrm{G}(k,N)) : \rho \mapsto \mathrm{ch}(E_\rho)$$

defined as follows. Denote by $E_\rho \to \mathrm{G}(k,N)$ the complex vector bundle associated to a representation $\rho : \mathrm{U}(k) \to \mathrm{Aut}(V)$ via $E_\rho = P \times_\rho V$. In this formula $P \to \mathrm{G}(k,N)$ denotes the principal $\mathrm{U}(k)$-bundle of unitary k-frames discussed above. The Chern character of a vector bundle E can be obtained from the Chern classes $c_j(E)$ via the formula

$$\mathrm{ch}(\xi) = \exp\left(\mathrm{trace}\left(\frac{i\xi}{2\pi}\right)\right), \qquad \det\left(\lambda\mathbb{1} - \frac{i\xi}{2\pi}\right) = \sum_{j=0}^{n} c_j(\xi)\lambda^{n-j}$$

where $n = \operatorname{rank} E$ and $\xi \in \mathfrak{u}(n)$. Alternatively, evaluate the power series ch on the curvature of a connection on E to obtain a differential form representing the class $\operatorname{ch}(E)$. This is the contents of Chern-Weil theory and we refer to Milnor–Stasheff [55] for details. The formulae

$$\operatorname{ch}(E \oplus F) = \operatorname{ch}(E) + \operatorname{ch}(F), \qquad \operatorname{ch}(E \otimes F) = \operatorname{ch}(E)\operatorname{ch}(F)$$

show that the Chern character determines an algebra homomorphism. This becomes in fact an isomorphism if we replace the finite dimensional Grassmannian by the limit of $G(k, N)$ as N tends to ∞. This limit is in fact the classifying space of the unitary group $U(k)$. If we consider only a fixed Grassmannian $G(k, N)$ then it is natural to restrict to a suitable subalgebra $\mathcal{R}_{k,N} \subset \mathcal{R}_k$.

Now there is a deformed product structure on the algebra $\mathcal{R}_{k,N}$ which is similar to quantum cohomology. This deformed product can roughly be described as follows. Given a compact oriented Riemann surface Σ of genus g we consider the moduli space M_Σ of flat $U(k)$-connections over Σ. This is a finite dimensional Kähler manifold (with singularities). It can be described as a quotient

$$M_\Sigma = \frac{\mathcal{A}_{\mathrm{flat}}(\Sigma)}{\mathcal{G}(\Sigma)}$$

of flat $U(k)$-connections $A \in \Omega^1(\Sigma, \mathfrak{u}(k))$ divided by the action of the gauge group $\mathcal{G}(\Sigma) = \operatorname{Map}(\Sigma, U(k))$. There is a natural holomorphic line bundle

$$L \to M_\Sigma$$

whose curvature is the symplectic form on M_Σ. We are interested in the dimension of the space $H^0(M_\Sigma, L^{\otimes \ell})$ of holomorphic sections of the ℓ-th power of this bundle. More generally, given marked points $z_1, \ldots, z_p \in \Sigma$ and representations $\rho_1, \ldots, \rho_p$ of $U(k)$, we can construct a representation

$$\rho : \mathcal{G}(\Sigma) \to V = V_1 \otimes \cdots \otimes V_p, \qquad \rho(u) = \rho_1(u(z_1)) \otimes \cdots \otimes \rho_p(u(z_p))$$

of the gauge group and form the vector bundle

$$E_\ell(\rho_1, \ldots, \rho_p) = L^{\otimes \ell} \otimes \mathcal{A}_{\mathrm{flat}}(\Sigma) \times_\rho V$$

over M_Σ. Now define the numbers

$$N_\ell(\Sigma; \rho_1, \ldots, \rho_p) = \dim H^0(M_\Sigma, E_\ell(\rho_1, \ldots, \rho_p)).$$

To give a precise meaning to these numbers one must define the space of holomorphic sections in the right way. This is particularly apparent in the case $\Sigma = S^2$ since in this case the moduli space of flat connections is just a point. However, this is a point with nontrivial isotropy subgroup and the spaces of holomorphic sections will be nontrivial. They should in fact be defined as global holomorphic maps $\mathcal{A}(\Sigma) \to \mathbb{C}$ on the space of all connections which are invariant under a suitable action of the gauge group $\mathcal{G}(\Sigma)$. Even if this programme has been carried out the above formula for $N_\ell(\Sigma; \rho_1, \ldots, \rho_p)$ is only correct if the higher cohomology groups vanish and should otherwise be replaced by a suitable Euler characteristic. Another difficulty arises from the presence of a $U(1)$ factor in $U(k)$. As a result we must specify a pair of integers (ℓ, m) (called the **level**) to characterize the required line bundle over M_Σ rather than just taking the ℓ-fold tensor product of the given bundle L. This would give rise to invariants $N_{\ell,m}$. In the following we shall discard the choice of the level in the notation and simply write $N(\Sigma; \rho_1, \ldots, \rho_p)$ instead of $N_\ell(\Sigma; \rho_1, \ldots, \rho_p)$.

The numbers $N(\Sigma; \rho_1, \ldots, \rho_p)$ satisfy the Verlinde gluing rules. One such rule is of the form

$$N(\Sigma_{g+1}; \rho_1, \ldots, \rho_p) = \sum_\rho N(\Sigma_g; \rho_1, \ldots, \rho_p, \rho, \rho^*)$$

where the sum is over a basis of $\mathcal{R}_{k,N}$. Here the subalgebra $\mathcal{R}_{k,N} \subset \mathcal{R}_k$ must be chosen an accordance with the level at which the invariants $N(\Sigma; \rho_1, \ldots, \rho_p)$ are defined. For example, in the slightly simpler context of the group $\mathrm{SU}(k)$, if the invariants are defined in terms of the ℓ-the power of the canonical line bundle $L \to M_\Sigma$, then the algebra should be chosen with generators $s^n(\mathbb{C}^k)$ (the symmetric powers of $\mathbb{C}^k$) up to order $|n| \leq \ell$. According to Witten, the right level to choose in connection with the Grassmannian $\mathrm{G}(k, N)$ is $\ell = N - k$, and the additional $\mathrm{U}(1)$ factor should be treated at the level $m = N$. Henceforth this will be our choice for the subalgebra $\mathcal{R}_{k,N} \subset \mathcal{R}_k$ and for the corresponding line bundle in the definition of the invariants.

A similar gluing rule takes the form

$$N(S^2; \rho_1, \rho_2, \rho_3, \rho_4) = \sum_\rho N(S^2; \rho_1, \rho_2, \rho^*) N(S^2; \rho, \rho_3, \rho_4). \qquad (8.16)$$

This is reminiscent of the composition rules for the Gromov-Witten invariants and can in fact be interpreted in terms of a deformed product structure on the algebra $\mathcal{R}_{k,N}$. The Verlinde product structure is defined by

$$\rho_\alpha * \rho_\beta = \sum_\gamma N(S^2; \rho_\alpha, \rho_\beta, \rho_\gamma^*) \rho_\gamma. \qquad (8.17)$$

The gluing rule (8.16) asserts that this product is associative. Now in [87] Witten conjectured that there should be an isomorphism

$$R_{k,N} \to QH^*(\mathrm{G}(k, N)) : \rho \mapsto \mathrm{qch}(E_\rho)$$

which maps the Verlinde product structure onto the quantum cup-product structure of the Grassmannian. This isomorphism should be a kind of quantum deformation of the Chern character where the deformation involves lower order terms. It is perhaps not unreasonable to expect that the quantum Chern character of a complex line bundle $L_\rho \to \mathrm{G}(k, N)$ should be the quantum exponential of the first Chern class (i.e. the exponential with multiplication replaced by the quantum cup-product). The existence of this isomorphism is a surprising and beautiful conjecture which should have interesting consequences. Moreover, there is a generalized version of this conjecture for Riemann surfaces of higher genus with marked points if one considers the corresponding Gromov-Witten invariants Ψ. For more details the reader may wish to consult [87].

8.5 The Gromov-Witten potential

In a recent paper [35] Kontsevich and Manin describe various remarkable structures related to the associativity rule and Frobenius algebra structure of quantum cohomology. Another discussion of these developments can be found in Ruan–Tian [67]. To describe these structures in their full generality requires the formalism of supermanifolds. In order to avoid these additional technicalities we shall consider in this section only the even dimensional part of the quantum cohomology groups, with complex coefficients, and denote it by

$$\mathcal{H} = H^{\mathrm{ev}}(M,\mathbb{C}) = \bigoplus_i H^{2i}(M,\mathbb{C}).$$

The Poincaré duality pairing with complex coefficients is the Hermitian form

$$\langle a, b \rangle = \int_M \overline{a} \cup b \tag{8.18}$$

for $a, b \in \mathcal{H}$. This form is complex linear in the second variable, complex anti-linear in the first variable and, because we consider only even classes, it satisfies

$$\langle a, b \rangle = \overline{\langle b, a \rangle}.$$

Note that the restriction of this pairing to integral classes agrees with the one defined in Section 8.1. It need not be positive definite. The extension of the quantum cup product to complex coefficients is given by the same formula as before, but with q set equal to $1 \in \mathbb{C}$. Thus

$$\langle a * b, c \rangle = \sum_A \Phi_A(\overline{\alpha}, \overline{\beta}, \gamma)$$

where $\alpha = \mathrm{PD}(a)$, $\beta = \mathrm{PD}(b)$, and $\gamma = \mathrm{PD}(c)$. Here we think of Φ_A as being complex linear in all variables and denote by $\overline{\alpha}$ complex conjugation. Observe that if a, b, c have pure degree – that is, if they each belong to some space $H^i(M;\mathbb{C})$ – the only classes A which contribute nontrivially to this sum are those for which $\deg(a) + \deg(b) + \deg(c) = 2n + 2c_1(A)$. In general, we must interpret this formula for $\langle a * b, c \rangle$ as

$$\sum_{i,j,k} \sum_A \Phi_A(\overline{\alpha}_i, \overline{\beta}_j, \gamma_k),$$

where $\sum \alpha_i$ is the decomposition of α into terms of pure degree, and similarly for $\sum \beta_j, \sum \gamma_k$.

With this this notation $\mathcal{H}$ is a complex Frobenius algebra in the sense that

$$\langle a * \overline{b}, c \rangle = \langle a, b * c \rangle.$$

The restriction to even classes suffices for many applications such as flag manifolds and Grassmannians.

Following Dubrovin [17] we can interpret the quantum cup product

$$\mathcal{H} \times \mathcal{H} \to \mathcal{H} : (a, b) \mapsto a * b$$

as a connection on the tangent bundle of $\mathcal{H}$ given by

$$\nabla_Y X(a) = dX(a)Y(a) + iX(a) * Y(a) \tag{8.19}$$

for $a \in \mathcal{H}$ and two vector fields $X, Y : \mathcal{H} \to \mathcal{H}$. (Here we identify all the tangent spaces of $\mathcal{H}$ with $\mathcal{H}$ in the obvious way, so that the derivative $dX(a)$ of the map X at a becomes a linear map of $\mathcal{H}$ to itself.) The commutativity of the quantum product can now be interpreted as the vanishing of the torsion, associativity can be interpreted as vanishing of the curvature and the Frobenius condition $\langle a * \bar{b}, c \rangle = \langle a, b * c \rangle$ means that ∇ is compatible with the Hermitian structure.

Lemma 8.5.1 (i) *The connection (8.19) is torsion free, i.e.*

$$[X, Y] = \nabla_Y X - \nabla_X Y.$$

(ii) *The connection (8.19) is compatible with the Hermitian structure, i.e.*

$$d\langle Y, Z \rangle \cdot X = \langle \nabla_{\overline{X}} Y, Z \rangle + \langle Y, \nabla_X Z \rangle$$

(iii) *The connection (8.19) is flat, i.e.*

$$\nabla_X \nabla_Y Z - \nabla_Y \nabla_X Z + \nabla_{[X,Y]} Z = 0.$$

(iv) *The connection (8.19) satisfies*

$$\nabla_X \mathbb{1} = iX.$$

Proof: Our sign convention for the Lie bracket is

$$[X, Y] = dX \cdot Y - dY \cdot X$$

where $dX(a)$ denotes the differential of the map $X : \mathcal{H} \to \mathcal{H}$ and should be thought of as a linear transformation of $\mathcal{H}$ which takes Y to $dX \cdot Y$. The first statement is now obvious. The second statement follows from the Frobenius condition by direct calculation:

$$\begin{aligned}
d\langle Y, Z \rangle \cdot X &= \langle dY \cdot \overline{X}, Z \rangle + \langle Y, dZ \cdot X \rangle \\
&= \langle \nabla_{\overline{X}} Y - iY * \overline{X}, Z \rangle + \langle Y, \nabla_X Z - iZ * X \rangle \\
&= \langle \nabla_X Y, Z \rangle + \langle Y, \nabla_X Z \rangle.
\end{aligned}$$

To prove flatness note that

$$\begin{aligned}
\nabla_X \nabla_Y Z &= \nabla_X (dZ \cdot Y + iZ * Y) \\
&= d(dZ \cdot Y + iZ * Y) \cdot X + i(dZ \cdot Y + iZ * Y) * X \\
&= d^2 Z(Y, X) + dZ \cdot dY \cdot X + iZ * (dY \cdot X) \\
&\quad + i(dZ \cdot X) * Y + i(dZ \cdot Y) * X - (Z * Y) * X \\
&= \nabla_{dY \cdot X} Z + d^2 Z(X, Y) - Z * (Y * X) \\
&\quad + i(dZ \cdot X) * Y + i(dZ \cdot Y) * X.
\end{aligned}$$

The fourth statement is obvious. $\square$

The above connection is of a very special form because the 1-form

$$A : T\mathcal{H} \to \mathrm{End}(\mathcal{H})$$

given by

$$A_a(x)y = ix * y$$

is constant and does not depend on the base point a. A general connection 1-form $A \in \Omega^1(\mathcal{H}, \mathrm{End}(\mathcal{H}))$ can be interpreted as a family of products

$$\mathcal{H} \times \mathcal{H} \to \mathcal{H} : (x, y) \mapsto x *_a y, \tag{8.20}$$

parametrized by the elements of $\mathcal{H}$ itself, via the formula

$$A_a(x)y = ix *_a y \tag{8.21}$$

for $a \in \mathcal{H}$ and $x, y \in T_a\mathcal{H} = \mathcal{H}$. The corresponding connection, when regarded as a differential operator $C^\infty(\mathcal{H}, \mathcal{H}) \to \Omega^1(\mathcal{H}, \mathcal{H})$ is given by

$$\nabla = d + A$$

or, more explicitly, by

$$\nabla_Y X(a) = dX(a)Y(a) + iX(a) *_a Y(a) \tag{8.22}$$

for two vector fields $X, Y : \mathcal{H} \to \mathcal{H}$. The properties of the **Dubrovin connection** (8.22) are related to the products (8.20) as follows. We assume here that the map $\mathcal{H} \times \mathcal{H} \times \mathcal{H} \to \mathcal{H} : (a, x, y) \mapsto x *_a y$ is holomorphic and complex bilinear in x and y. We shall consider the case where these products determine a family of Frobenius algebra structures, one on each tangent space $T_a\mathcal{H} = \mathcal{H}$.

Lemma 8.5.2 (i) *The connection (8.22) is torsion free if and only if the products (8.20) are commutative, i.e.*

$$x *_a y = y *_a x$$

for all $a, x, y \in \mathcal{H}$.

(ii) *Assume that the connection (8.22) is torsion free. Then it is compatible with the Hermitian structure if and only if the products (8.20) satisfy the Frobenius condition*

$$\langle \overline{x} * \overline{a}\overline{y}, z \rangle = \langle x, y *_a z \rangle$$

for all $a, x, y, z \in \mathcal{H}$.

(iii) *Assume that the connection (8.22) is Hermitian and torsion free. Then the 1-form (8.21) is closed if and only if there exists a holomorphic function $\mathcal{S} : \mathcal{H} \to \mathbb{C}$ such that*

$$\langle \overline{x} * \overline{a}\overline{y}, z \rangle = \partial^3 \mathcal{S}_a(x, y, z) \tag{8.23}$$

(iv) *Assume that the connection (8.22) is Hermitian and torsion free. Then the 1-form (8.21) satisfies*

$$A \wedge A = 0$$

if and only if the products (8.20) are all associative.

(v) *The connection (8.19) satisfies $\nabla_X \mathbb{1} = iX$ if and only if $\mathbb{1}$ is a unit for all the products (8.20).*

Proof: The first two statements are proved as in Lemma 8.5.1. To prove (iii) note that the function

$$\phi_a(x, y, z) = \langle \overline{x} * \overline{a}\overline{y}, z \rangle$$

is symmetric in x, y, z. Moreover, a simple calculation shows that its derivative

$$\psi_a(w, x, y, z) = \left. \frac{d}{dt} \right|_{t=0} \phi_{a+tw}(x, y, z)$$

is symmetric in w, x, y, z if and only if the connection 1-form A given by (8.21) is closed. Now the symmetry of ψ_a is equivalent to the existence of a holomorphic function $S : \mathcal{H} \to \mathbb{C}$ such that

$$\phi_a(x, y, z) = \partial^3 S_a(x, y, z).$$

In fact, an explicit formula for S is given by

$$S(a) = \frac{1}{2} \int_0^1 (1 - t)^2 \phi_{ta}(a, a, a)\, dt.$$

The statements (iv) and (v) are obvious. $\qquad\qquad\square$

Remark 8.5.3 The proof of statement (iii) in the previous lemma can be formulated more explicitly in terms of a complex basis $e_0, \ldots, e_m$ of the cohomology $\mathcal{H} = H^{\mathrm{ev}}(M, \mathbb{C})$. Then $\mathcal{H}$ can be identified with $\mathbb{C}^{m+1}$ via the isomorphism $\mathbb{C}^{m+1} \to \mathcal{H} : x \mapsto \sum_i x^i e_i$. Define the functions $A_{ijk} : \mathbb{C}^{m+1} \to \mathbb{C}$ by

$$A_{ijk}(x) = \langle \overline{e}_i * \overline{a}\overline{e}_j, e_k \rangle, \qquad a = \sum_i x^i e_i.$$

These functions are holomorphic by assumption. They are symmetric under permutations of i, j, k if and only if the products (8.20) are symmetric and satisfy the Frobenius condition of (ii) in Lemma 8.5.2. If this holds then there exists a holomorphic function $S : \mathbb{C}^{m+1} \to \mathbb{C}$ such that

$$\frac{\partial^3 S}{\partial x^i \partial x^j \partial x^k} = A_{ijk}$$

if and only if the derivatives $\partial_\ell A_{ijk}$ are symmetric under permutations of i, j, k, ℓ. An explicit formula for S is given by

$$S(x) = \frac{1}{2} \int_0^1 (1 - t)^2 \sum_{i,j,k} A_{ijk}(tx) x^i x^j x^k\, dt.$$

In [35] all the conditions of Lemma 8.5.2 are formulated in local coordinates. In particular one can introduce a matrix

$$g_{ij} = g_{ji} = \langle \overline{e}_i, e_j \rangle$$

which represents the metric and obtain

$$e_i *_a e_j = \sum_\ell A_{ij}^\ell(x)e_\ell, \qquad a = \sum_i x^i e_i$$

where

$$A_{ij}^\ell(x) = \sum_k g^{\ell k} A_{ijk}$$

and g^{ij} represents the inverse matrix of g_{ij}. $\qquad\qquad\square$

A holomorphic function

$$\mathcal{S} : \mathcal{H} \to \mathbb{C}$$

is called a **potential function** for the Dubrovin connection (8.22) if its third derivatives satisfy the equation (8.23). In view of Lemma 8.5.2 such a function exists if and only if the connection is Hermitian and torsion free and the connection 1-form is closed. Conversely, if the potential function $\mathcal{S} : \mathcal{H} \to \mathbb{C}$ is given and and the products (8.20) are defined by (8.23) then these products are automatically commutative and satisfy the Frobenius condition. Moreover, $\mathbb{1}$ is a unit if and only if the function $\mathcal{S}$ satisfies

$$\partial^3 \mathcal{S}_a(\mathbb{1}, x, y) = \langle \overline{x}, y \rangle. \tag{8.24}$$

Associativity translates into the remarkable system of quadratic third order partial differential equations

$$\sum_i \partial^3 \mathcal{S}_a(w, x, e_i)\partial^3 \mathcal{S}_a(f_i, y, z) = \sum_i \partial^3 \mathcal{S}_a(w, z, e_i)\partial^3 \mathcal{S}_a(f_i, x, y) \tag{8.25}$$

where e_i denotes a basis of $\mathcal{H}$ and f_j denotes the dual basis with respect to the Hermitian form (8.18), i.e.

$$\langle \overline{f}_j, e_i \rangle = \delta_{ij}.$$

Equation (8.25) is called the **WDVV-equation** (as in Witten-Dijkgraaf-Verlinde-Verlinde).

Remark 8.5.4 In the notation of Remark 8.5.3 with $e_0 = \mathbb{1}$ the condition (8.24) takes the form

$$\frac{\partial^3 \mathcal{S}}{\partial x^0 \partial x^i \partial x^j} = g_{ij}.$$

and the WDVV-equation can be written as

$$\sum_{\nu,\mu} \frac{\partial^3 \mathcal{S}}{\partial x^i \partial x^j \partial x^\nu} g^{\nu\mu} \frac{\partial^3 \mathcal{S}}{\partial x^\mu \partial x^k \partial x^\ell} = \sum_{\nu,\mu} \frac{\partial^3 \mathcal{S}}{\partial x^i \partial x^\ell \partial x^\nu} g^{\nu\mu} \frac{\partial^3 \mathcal{S}}{\partial x^\mu \partial x^j \partial x^k}$$

for all i, j, k, ℓ. $\qquad\qquad\square$

Lemma 8.5.5 *Let $S : \mathcal{H} \to \mathbb{C}$ be a holomorphic function which satisfies (8.24) and let the quantum multiplication $x *_a y$ be defined by (8.20) for $a \in \mathcal{H}$. Then the following are equivalent.*

(i) *The products $x *_a y$ are associative.*

(ii) *The Dubrovin connection (8.22) is flat.*

(iii) *The potential function S satisfies the WDVV-equation (8.25).*

Proof: The equivalence of (i) and (ii) follows from the fact that the connection 1-form A, given by (8.21), is closed and, in view of Lemma 8.5.2, the condition $A \wedge A = 0$ is equivalent to associativity. Since the curvature of ∇ is the endomorphism valued 2-form

$$F_A = dA + A \wedge A$$

we obtain in fact that associativity is equivalent to the condition that all the connections

$$\nabla_\lambda = d + \lambda A$$

with $\lambda \in \mathbb{R}$ are flat.

To prove the equivalence of (i) and (iii) note that $y = \sum_i \langle \overline{f}_i, y \rangle e_i$ and hence

$$\langle x, y \rangle = \sum_i \langle x, e_i \rangle \langle \overline{f}_i, y \rangle$$

for all $x, y \in \mathcal{H}$. This implies

$$\sum_i \partial^3 S_a(w, x, e_i) \partial^3 S_a(f_i, y, z) = \sum_i \langle \overline{w} * \overline{a}\overline{x}, e_i \rangle \langle \overline{f}_i, y *_a z \rangle$$
$$= \langle \overline{w} * \overline{a}\overline{x}, y *_a z \rangle$$
$$= \langle \overline{w}, x *_a (y *_a z) \rangle$$

and therefore associativity is equivalent to the condition that the left hand side of (8.25) is symmetric under permutations of x, y, and z. This proves the lemma.
$\square$

If we consider the quantum deformation of the cup-product which is independent of a then the corresponding potential function S is a cubic polynomial. By definition of the quantum cup product this cubic polynomial is given by

$$S_3(a) = \frac{1}{3!} \sum_A \Phi_A(\alpha, \alpha, \alpha)$$

where $\alpha = \mathrm{PD}(a)$. It is another remarkable observation of Witten that the function

$$S(a) = \sum_{p \geq 3} \frac{1}{p!} \sum_A \Phi_{A,p}(\alpha, \ldots, \alpha) \tag{8.26}$$

with $\alpha = \mathrm{PD}(a)$ should be a solution of the WDVV-equation (8.25). As before, the class a may be a sum of terms of different degrees, i.e. $a = \sum_i a_i$ where $\deg(a_i) = d_i$. In this case the term $\Phi_{A,p}(\alpha, \ldots, \alpha)$ should be interpreted as the sum of the terms $\Phi_{A,p}(\alpha_{i_1}, \ldots, \alpha_{i_p})$ over all multi-indices $(i_1, \ldots, i_p)$. The only nonzero terms in this sum are those where A satisfies the dimension condition

$$2c_1(A) = \sum_{j=1}^{p}(\deg(a_{i_j}) - 2) + 6 - 2n. \tag{8.27}$$

This condition is equivalent to (7.2) and guarantees finiteness of the second sum for fixed p (in the monotone case) but there are infinitely many possibly nonzero terms corresponding to increasing values of p. As a result there is a nontrivial convergence problem, even in the monotone case. Note also the interesting case of Calabi-Yau manifolds with $n = 3$ and $c_1 = 0$. In this case the right hand side of (8.27) is zero for $a \in H^2(M)$, and so the dimensional condition is satisfied for all classes A. So in this case the sum on the right is infinite, even for fixed p, and we must multiply the terms by a factor $e^{-t\omega(A)}$ to have any hope of convergence.

The fact that the third order term in $\mathcal{S}$ satisfies the WDVV-equation is precisely the formula of Corollary 8.2.6 and is, of course, equivalent to the associativity law proved above. The proof that the whole function (8.26), if the series converges, is a solution of (8.25) requires a gluing argument for J-holomorphic curves which, as we now explain, involves the mixed invariants of Ruan and Tian [67] or equivalently the higher codimension classes of Kontsevich and Manin [35].

Notice that the third derivative of the Gromov-Witten potential (8.26) is given by

$$\partial^3 \mathcal{S}_a(x, y, z) = \sum_{p \geq 3} \frac{1}{(p-3)!} \sum_A \Phi_{A,p}(\xi, \eta, \zeta, \alpha, \ldots, \alpha)$$

where $\alpha = \mathrm{PD}(a)$, $\xi = \mathrm{PD}(x)$, $\eta = \mathrm{PD}(y)$, $\zeta = \mathrm{PD}(z)$. If the classes a, x, y, and z are all of pure degree then the only classes A which contribute nontrivially to the sum are those which satisfy the dimension condition

$$2c_1(A) = (p-3)(\deg(a) - 2) + \deg(x) + \deg(y) + \deg(z) - 2n.$$

(In general this condition has to be appropriately modified.) The corresponding product of $x \in H^k(M)$ and $y \in H^\ell(M)$ (with k and ℓ even) is given by

$$x *_a y = \sum_p \frac{1}{(p-3)!} \sum_A (x *_a y)_{A,p}$$

where $(x *_a y)_{A,p} \in QH^m(M)$ is defined by

$$\int_\zeta (x *_a y)_{A,p} = \Phi_{A,p}(\xi, \eta, \zeta, \alpha, \ldots, \alpha)$$

for $\zeta \in H_m(M)$ with $m = k + \ell + (p-3)(\deg(a) - 2) - 2c_1(A)$.

Now one can show as in Section 8.2 that the WDVV-equation for the Gromov-Witten potential translates into the composition rule

$$\Psi_{A,p}(\theta, \xi, \eta, \zeta; \alpha, \ldots, \alpha) = \sum_{q=3}^{p-1} \binom{p-4}{q-3} \tag{8.28}$$

$$\sum_{B,i} \Phi_{B,q}(\theta, \xi, \varepsilon_i, \alpha, \ldots, \alpha)\Phi_{A-B,p-q+2}(\phi_i, \eta, \zeta, \alpha, \ldots, \alpha)$$

where $\Psi_{A,p}$ denotes the so-called *mixed invariant* defined with 4 marked points on $\mathbb{C}P^1$ and arbitrary intersection points otherwise, and where ε_i, ϕ_i are as in Lemma 8.2.4. The equation (8.28) is correctly stated for even homology classes. In the general case there is a similar equation which involves signs corresponding to permutations (see Kontsevich–Manin [35] or Ruan Tian [67]).

The proof of (8.28) involves a gluing argument as in Section 8.2 and Appendix A. One interprets the expression on the right in terms of intersecting pairs of J-holomorphic curves, which represent the classes B and $A - B$ and intersect the appropriate homology classes. The analytical details are no different than in the case $p = 4$ of 4 intersection points. An alternative proof is given in [67].

Example 8.5.6 In the case $M = \mathbb{C}P^n$ the cohomology ring $\mathcal{H}$ can be naturally identified with $\mathbb{C}^{n+1}$ where the point $z = (z_0, \ldots, z_n)$ corresponds to the cohomology class

$$a = z_0 \mathbb{1} + z_1 p + z_2 p^2 + \cdots + z_n p^n.$$

We compute the third order term $\mathcal{S}_3$ of the Gromov-Witten potential $\mathcal{S}$. In view of Example 8.1.6, it is not hard to check that this term is given by

$$\mathcal{S}_3(z) = \frac{1}{6} \left(\sum_{i+j+k=n} z_i z_j z_k + \sum_{i+j+k=2n+1} z_i z_j z_k \right). \qquad \Box$$

Example 8.5.7 In the case $M = \mathbb{C}P^1$ the system (8.25) imposes no condition at all on the function $\mathcal{S}$ and the Gromov-Witten potential is given by

$$\mathcal{S}(z_0, z_1) = \tfrac{1}{2} z_0^2 z_1 + e^{z_1} - 1 - z_1 - \tfrac{1}{2} z_1^2$$

(cf. Kontsevich and Manin [35]). This is equivalent to the obvious fact that $\Phi_{A,p}(\mathrm{pt}, \ldots, \mathrm{pt}) = 1$ for $A = [\mathbb{C}P^1]$ and any p while all invariants with $A = k[\mathbb{C}P^1]$, $k \geq 2$, are zero. $\qquad \Box$

Example 8.5.8 In the case $M = \mathbb{C}P^2$ the equation (8.24) takes the form

$$\frac{\partial^3 \mathcal{S}}{\partial z_0 \partial z_0 \partial z_2} = 1, \qquad \frac{\partial^3 \mathcal{S}}{\partial z_0 \partial z_1 \partial z_1} = 1,$$

and all other third derivatives involving z_0 vanish. Moreover, it was observed by Kontsevich and Manin in [35] that in this case the system (8.25) is equivalent to the single differential equation

$$\frac{\partial^3 \mathcal{S}}{\partial z_2 \partial z_2 \partial z_2} + \frac{\partial^3 \mathcal{S}}{\partial z_1 \partial z_1 \partial z_1} \frac{\partial^3 \mathcal{S}}{\partial z_1 \partial z_2 \partial z_2} = \left(\frac{\partial^3 \mathcal{S}}{\partial z_1 \partial z_1 \partial z_2} \right)^2.$$

This corresponds to the case $i = j = 1$, $k = \ell = 2$ in Remark 8.5.4. Now let

$$N(d) = \Phi_{dL, 3d-1}(\mathrm{pt}, \ldots, \mathrm{pt})$$

be the number of rational curves of degree d in $\mathbb{C}P^2$ intersecting $p = 3d - 1$ generic points. This number is well defined because the condition (8.27) is satisfied with

$\alpha = [\mathrm{pt}]$, $\deg(a) = 4$, $A = dL$, $p = 3d - 1$, and $n = 2$. The Gromov-Witten potential is given by[4]

$$\mathcal{S}(z) = \tfrac{1}{2}(z_0 z_1^2 + z_0^2 z_2) + \sum_{d=1}^{\infty} N(d) \frac{z_2^{3d-1}}{(3d-1)!} e^{dz_1}$$

and the above differential equation is equivalent to the recursive formula

$$N(d) = \sum_{k+\ell=d} N(k)N(\ell)k^2\ell \left(\ell \begin{pmatrix} 3d - 4 \\ 3k - 2 \end{pmatrix} - k \begin{pmatrix} 3d - 4 \\ 3k - 1 \end{pmatrix} \right)$$

for $d \geq 2$ with $N(1) = 1$ (cf. Kontsevich and Manin [35]). Observe that in this case the recursion formula follows directly from the composition rule (8.28) and so it holds regardless of the convergence of $\mathcal{S}$. It also uniquely determines the numbers $N(d)$ and the Gromov-Witten potential $\mathcal{S}$. The first few values of $N(d)$ are $N(2) = 1$, $N(3) = 12$, $N(4) = 620$, $N(5) = 87304$, $N(6) = 26312976$. $\qquad\square$

Remark 8.5.9 (i) If we include the odd-dimensional cohomology groups in the above discussion then the commutativity and the action of the permutation group involves signs and this leads naturally to the notion of a supermanifold. The signs will also be involved in the correct definition of the Gromov-Witten potential (8.26). We shall not discuss this extension here and refer the interested reader to [17] and [35].

(ii) The definition of the Gromov-Witten potential can also be extended to manifolds which are not monotone. However this would involve exponents of the form $e^{-t\omega(A)}$ and one has to solve a nontrivial convergence problem as explained in the next chapter.

$\qquad\square$

---

[4]This formula contains a second order term $\frac{1}{2}z_2{}^2$ which corresponds to the formula $\Phi_L(\mathrm{pt}, \mathrm{pt}) = 1$ and, of course, does not affect the third derivatives.

Chapter 9

Novikov Rings and Calabi-Yau Manifolds

The goal of this chapter is to extend the definition of the Gromov-Witten invariants to all weakly monotone symplectic manifolds. An application to quantum cohomology for Calabi-Yau manifolds is given in Section 9.3. Recall from Remark 5.1.4 that (M, ω) is weakly monotone if either it is monotone, or the first Chern class vanishes over $\pi_2(M)$, or the minimal Chern number is $N \geq n - 2$. So far we have only dealt with the monotone case with minimal Chern number $N \geq 2$.

In general, if the manifold M is only weakly monotone, then the dimension condition (8.3) on $c_1(A)$ will no longer guarantee that the energy of the curve A is uniformly bounded and hence the sum in (8.2) will no longer be finite. In this case the deformed cup product can be defined by

$$\langle a *_t b, c \rangle = \sum_{i,j,k} \sum_A \Phi_A(\alpha_i, \beta_j, \gamma_k) e^{-t\omega(A)} \tag{9.1}$$

for $c \in QH^*(M)$. As before, the sum should run over all quadruples (i, j, k, A) which satisfy $N(i + j + k) + c_1(A) = 0$. This is a beautiful formula and shows how the quantum deformed cup product is actually a *deformation* of the ordinary cup product for large t. However, as is often the case with beautiful formulae, there are several problems to overcome in order to make this rigorous. The first, and not so serious, problem is the presence of multiply covered curves of Chern number zero. We shall see in Section 9.1 how to circumvent this difficulty. The second, and rather more serious, problem is the question of convergence. We must find a uniform exponential bound on the invariants $\Phi_A(\alpha, \beta, \gamma)$ in terms of $e^{c\omega(A)}$ for some constant c. That such a bound should exist has been conjectured by physicists in the case of Calabi-Yau manifolds (cf. [8], [9]). This convergence problem is related to enumerative problems in algebraic geometry. Without solving this convergence question one can work instead with formal sums and this leads naturally to the Novikov ring. We shall discuss this approach in Section 9.2.

141

9.1   Multiply-covered curves

We now consider the problem of extending the definition of $\Phi_A(\alpha, \beta, \gamma)$ to the case where $A = mB$ is a nontrivial multiple of a class B with $c_1(B) = 0$. For such classes there is first of all the problem of proving that the number of simple curves representing A (and intersecting α, β, γ) is finite under the appropriate conditions on the dimensions of the homology classes and on the regularity of the almost complex structure J. Secondly, there is the question of how to take account of the multiply-covered curves which represent the class A. Moreover, in the proof of associativity there is the difficulty of extending the definition of $\Psi_A(\alpha, \beta, \gamma, \delta)$ to cases in which condition (JA_4) is not satisfied. In this section, we shall outline a way to get around these difficulties.

As we have seen in Chapter 6, when $A = mB$ is a nontrivial multiple of a class B with $c_1(B) > 0$ and $m > 1$, then the multiply covered B-curves form part of the boundary of the image $X(A, J)$ of the evaluation map and so do not contribute to Φ_A. However, if $c_1(B) = 0$, the moduli spaces $\mathcal{M}(A, J)$ and $\mathcal{M}(B, J)$ have the same dimension and the statements in Theorems 5.2.1 and 5.3.1 break down. To illustrate the problems which occur in this situation, let us consider the case when M is a 6-dimensional manifold with $c_1 = 0$. Then the dimension of $\mathcal{M}(A, J)/G$ is zero for all A and this implies that for a generic almot complex structure J all (unparametrized) A-curves are isolated. However, if $A = kB$, what we have proved so far does not imply that there are only finitely many A-curves, since there could be a sequence of A-curves which converge to a multiply covered B-curve $u : \mathbb{C}P^1 \to M$. This will not happen if J is integrable near the curve $C_B = u(\mathbb{C}P^1)$ and if u is an embedding. In this case it follows from the surjectivity of the linearized $\bar\partial$-operator D_u that the normal bundle of C_B must have type $(-1, -1)$, i.e. it decomposes into a sum of holomorphic line bundles each of which has Chern number -1. This implies that the pull-back of the normal bundle by a map of degree k has no sections and hence, by results on complex geometry, that there are no A-curves near C_B. However, it is not clear how to extend this argument to the case of non-integrable J. Moreover, in order to show that Φ_A and Ψ_A are well defined, one would have to show that such a sequence of A curves converging to a multiply covered B-curve can only occur for a set of almost complex structures J of codimension at least 2. Even if the above problems are solved, so that one does have a well-defined finite number of simple A-curves, the problem remains of how to count those A-curves which are multiple covers of some B-curve.

As noted by Ruan, one can get around both these problems by using Gromov's trick of considering the graphs

$$\widehat{u} : \mathbb{C}P^1 \to \widehat{M} = \mathbb{C}P^1 \times M, \qquad \widehat{u}(z) = (z, u(z)),$$

of J-holomorphic curves $u : \mathbb{C}P^1 \to M$. These curves are $\widehat{J}$-holomorphic with respect to the almost complex structure

$$\widehat{J} = i \times J$$

on $\widehat{M}$. Moreover, if J is ω-tame (or ω-compatible) then the product structure $\widehat{J}$ is $\widehat{\omega}$-tame (or $\widehat{\omega}$-compatible) where $\widehat{\omega} = \tau_0 \times \omega$ and τ_0 is the standard symplectic form on $\mathbb{C}P^1$ corresponding to the Fubini-Study metric. Conversely, every $\widehat{J}$-holomorphic curve with projection $\phi = \pi_1 \circ \widehat{u} : \mathbb{C}P^1 \to \mathbb{C}P^1$ of degree 1 is the graph of a J-holomorphic curve up to reparametrization. In [26], Gromov used this method in a more general context to interpret graphs of solutions $u : \mathbb{C}P^1 \to M$ of an inhomogeneous equation of the form $\bar{\partial}_J(u) = h$ as a $\widehat{J}_h$-holomorphic curve, where the complex structure $\widehat{J}_h$ depends on h.

Denote by A_0 the homology class $[\mathbb{C}P^1 \times \{\text{pt}\}] \in H_2(\widehat{M})$ and, given $A \in H_2(M)$, denote $\widehat{A} = A_0 + A$. Then, instead of counting A-curves in M, we can count $\widehat{A}$-curves in $\widehat{M}$. These curves cannot be multiply covered because $\widehat{A}$ is not a multiple class. However, because we must allow for a generic perturbation of the almost complex structure, we cannot in general work with the product structure $\widehat{J} = i \times J$ but with an almost complex structure nearby. (In fact, it suffices to consider structures of the form $\widehat{J}_h$ above, whose curves are graphs of the inhomogeneous equation $\bar{\partial}_J(u) = h$.)

For each $u \in \mathcal{M}(A, J)$ there is a 6-dimensional family of reparametrizations of the graph $\widehat{u}$. This corresponds to the formal relation between the dimensions of the moduli spaces for generic almost complex structures:

$$\dim \mathcal{M}(\widehat{A}, \widehat{J}) = \dim \mathcal{M}(A, J) + 6.$$

Observe further that each class $\alpha \in H_*(M)$ gives rise to an element in $H_*(\widehat{M})$ in two ways. There is the homology class of $\{z\} \times \alpha$ for $z \in \mathbb{C}P^1$ which has the same degree as α and also the class $\widehat{\alpha} = [\mathbb{C}P^1 \times \alpha]$ of degree 2 higher.

Lemma 9.1.1 *Assume that the manifold (M, ω) is weakly monotone.*

(i) *If the almost complex structure $J \in \mathcal{J}(M, \omega)$ is semi-positive then so is the product structure $\widehat{J} = i \times J \in \mathcal{J}(\widehat{M}, \widehat{\omega})$.*

(ii) *Assume that A is not a nontrivial multiple of a class B with $c_1(B) = 0$. Assume, moreover, that $J \in \mathcal{J}_+(M, \omega, K)$ for some $K > \omega(A)$ and is regular in the sense of Theorem 5.3.1. Then $\widehat{J}$ is also regular and*

$$\Phi_{A,J}(\alpha_1, \ldots, \alpha_p) = \Phi_{\widehat{A}, \widehat{J}}(\{z_1\} \times \alpha_1, \ldots, \{z_3\} \times \alpha_3, \widehat{\alpha}_4, \ldots, \widehat{\alpha}_p)$$

for any three distinct points $z_1, z_2, z_3 \in \mathbb{C}P^1$.

(iii) *If $A = mB$ is a nontrivial multiple of a class B assume that either $c_1(B) \geq 3$ or $p \leq 2m$. Moreover, assume that $J \in \mathcal{J}(M, \omega)$ is regular in the sense of Theorem 5.4.1 and that every J-holomorphic curve has Chern number at least 2. Then $\widehat{J} = i \times J$ has the same properties and*

$$\Psi_{A,J}(\alpha_1, \ldots, \alpha_p) = \Psi_{\widehat{A}, \widehat{J}}(\{z_1\} \times \alpha_1, \ldots, \{z_3\} \times \alpha_3, \widehat{\alpha}_4, \ldots, \widehat{\alpha}_p)$$

for any three distinct points $z_1, z_2, z_3 \in \mathbb{C}P^1$.

Proof: Exercise. $\qquad\square$

It is not hard to see that the right hand side in these identities is well-defined for all classes A and all $p \geq 3$ whenever M is weakly monotone and the almost complex structure $\widehat{J}$ is sufficiently close to a product structure. In this case, a sequence $\widehat{u}_\nu \in \mathcal{M}(\widehat{A}, \widehat{J})$ with $\widehat{u}_\nu(\zeta_j) \in \{z_j\} \times \alpha_j$ cannot converge (modulo bubbling) to a multiply covered curve. This is because the projection $\phi_\nu = \pi_1 \circ \widehat{u}_\nu$ satisfies $\phi_\nu(\zeta_j) = z_j$ and, in the case $\widehat{J} = i \times J$, is a holomorphic map of degree 1. These conditions determine ϕ_ν completely. In the general case, ϕ_ν satisfies a slightly perturbed inhomogeneous Cauchy-Riemann equation and the points z_j will also be perturbed because the cycles representing $\{z_j\} \times \alpha_j$ must be put in general position. But it is still impossible for ϕ_ν to converge to a multiply covered curve. Hence the proof of Theorem 5.4.1 shows that the codimension argument works for each class $\widehat{A}$. It follows that the invariants $\Phi_{\widehat{A},\widehat{J}}$ and $\Psi_{\widehat{A},\widehat{J}}$ are always well defined and do not depend on the choice of $\widehat{J}$, provided that this is sufficiently near a product structure. In view of Lemma 9.1.1 we can use these invariants to define Φ_A and Ψ_A for all classes $A \in H_2(M, \mathbb{Z})$ in a weakly monotone symplectic manifold.

This is the approach taken by Ruan and Tian in [67]. It allows us to extend the definition of quantum cohomology to all weakly monotone symplectic manifolds. Also the proof of associativity will be essentially the same as in Section 8.2. Of course, the problem of computing the deformed cup product will in general be nontrivial and is a topic for future research. We give one example at the end of this chapter.

9.2   Novikov rings

If the symplectic manifold M is not monotone then the set of J-holomorphic curves with given Chern number will in general not have uniformly bounded energy. One way to overcome this difficulty is to count J-holomorphic curves in a given homology class and this leads naturally to the Novikov ring $\Lambda = \Lambda_\omega$ associated to the homomorphism

$$\omega : \Gamma \to \mathbb{Z}.$$

Here $\Gamma \subset H_2(M)$ is the image of the Hurewicz homomorphism $\pi_2(M) \to H_2(M)$ and ω is the symplectic form on M. One should think of the Novikov ring as a completion of the group ring of Γ, which is associated to a grading of Γ induced by ω. Thus it is more like a Laurent ring than a polynomial ring. Because we have divided out by the torsion subgroup, Γ is isomorphic to $\mathbb{Z}^m$ for some m and below we shall give a description which depends on an explicit choice of this isomorphism. But we begin our discussion in a coordinate free setting.

An element of the Novikov ring can be thought of as a Fourier series of the form

$$\lambda = \sum_{A \in \Gamma} \lambda_A e^{2\pi i A}$$

where $\lambda_A \in \mathbb{Z}$. Alternatively, we could allow the λ_A to be in any principal ideal domain. The Novikov ring Λ_ω consists of all formal sums λ of this form, such that, for each $c > 0$, there are only finitely many nonzero coefficients λ_A with energy $\omega(A) \leq c$. In other words, the coefficients λ_A are subject to the finiteness condition

$$\#\{A \in \Gamma \,|\, \lambda_A \neq 0, \ \omega(A) \leq c\} < \infty$$

for every $c > 0$. Thus the Novikov ring is a kind of completion of the group ring of Γ. The ring structure is given by

$$\lambda * \mu = \sum_{A,B} \lambda_A \mu_B e^{2\pi i (A+B)}.$$

Thus

$$(\lambda * \mu)_A = \sum_B \lambda_{A-B}\mu_B.$$

It is a simple matter to check that the finiteness condition is preserved under this multiplication.

The Novikov ring carries a natural grading given by the first Chern class via

$$\deg(e^{2\pi iA}) = 2c_1(A).$$

If our symplectic manifold is monotone, this grading agrees up to a positive factor with the energy level $\omega(A)$ used in the finiteness condition. The monotonicity condition also implies that the homomorphism $\omega : \Gamma \to \mathbb{Z}$ has an $(m-1)$-dimensional kernel (if $\Gamma \cong \mathbb{Z}^m$). If the monotonicity condition is dropped then ω will in general not be an integral class. However, it does determine a homomorphism $\omega : H_2(M) \to \mathbb{R}$, which may be injective. This case may occur for example in Calabi-Yau manifolds where the first Chern class is zero. Note that in the case $c_1 = 0$ the Novikov ring is not graded.

An integral basis $A_1, \ldots, A_m$ of Γ determines an explicit isomorphism $\mathbb{Z}^m \to \Gamma$ which assigns to every integer vector $d \in \mathbb{Z}^m$ the homology class

$$A_d = d_1 A_1 + \cdots + d_m A_m.$$

Correspondingly, the vector d is assigned the *energy level* $\omega(A_d)$. We may in this case introduce the auxiliary variables $q_j = e^{2\pi iA_j}$ which we think of as multiplicative representatives of the classes A_j With this formal notation the elements of the Novikov ring are formal sums

$$\lambda = \sum_d \lambda_d q^d, \qquad q^d = q_1^{d_1} \cdots q_m^{d_m}.$$

The finiteness condition now becomes

$$\#\left\{d \in \mathbb{Z}^m \mid \lambda_d \neq 0,\ \omega(A_d) \leq c\right\} < \infty$$

and the grading is given by $\deg q^d = 2c_1(A_d)$. Note that the exponent d may have negative components.

Now consider the special case where $\Gamma = H_2(M)$. This occurs, for example, when M is simply connected. In this case the map $\mathbb{Z}^m \to H_2(M) : d \mapsto A_d$ is an isomorphism and the components $p_1, \ldots, p_m$ of the inverse isomorphism $p : H_2(M) \to \mathbb{Z}^m$ can be interpreted as cohomology classes in $H^2(M)$. They form a basis of $H^2(M)$ which is dual to the basis $A_1, \ldots, A_m$ of $H_2(M)$ since $p_i(A_j) = \delta_{ij}$. Moreover, in this case we may think of the auxiliary variables q_j which were introduced above as coordinates on the cohomology groups $H^2(M, \mathbb{C}/\mathbb{Z}) = H^2(M, \mathbb{C})/H^2(M, \mathbb{Z})$ given by

$$q_j(a) = e^{2\pi i \int_{A_j} a}$$

for $a \in H^2(M, \mathbb{C})$. This would give a rigorous meaning to the formal expression $q_j = e^{2\pi iA_j}$. Each coordinate function $q_j : H^2(M, \mathbb{C}/\mathbb{Z}) \to \mathbb{C}^*$ is a homomorphism with respect to the multiplicative structure of $\mathbb{C}^*$. We took essentially this approach when discussing flag manifolds, except that we considered only positive powers of the generators q_i.

Remark 9.2.1 (i) If $c_1 \neq 0$ we denote by $\Lambda_k \subset \Lambda$ the subset of all elements of degree k. Then Λ_0 is a ring, but, in general, multiplication changes the degree via the formula

$$\deg(\lambda * \mu) = \deg(\lambda) + \deg(\mu).$$

In other words, Λ_k is a module over Λ_0. Moreover, multiplication by any element of degree k provides a bijection $\Lambda_0 \to \Lambda_k$. Note that $\Lambda_k \neq \emptyset$ if and only if k is an integer multiple of $2N$ where N is the minimal Chern number of M (defined by $\langle c_1, \pi_2(M) \rangle = N\mathbb{Z}$).

(ii) If $\Gamma = \mathbb{Z}$ then Λ is the ring of Laurent series with integer coefficients. This is a principal ideal domain and if the coefficients are taken in a field then Λ is a field. These observations remain valid when the homomorphism $\omega : \Gamma \to \mathbb{R}$ is injective. (See for example [32].) In the case $\pi_2(M) = \mathbb{Z}$ it is interesting to note the difference between $c_1 = 0$ and $c_1 = [\omega]$. In both cases Λ_ω is the ring of Laurent series but if $c_1 = 0$ then this ring is not graded and in general we cannot exclude the possibility of infinitely many nonzero coefficients. This case appears for example when M is a quintic hypersurface in $\mathbb{C}P^4$ (see below).

(iii) Novikov first introduced a ring of the form Λ_ω in the context of his Morse theory for closed 1-forms (cf. [58]). In that case Γ is replaced by the fundamental group and the homomorphism $\pi_1(M) \to \mathbb{R}$ is induced by the closed 1-form.

(iv) As we shall see in Chapter 10 below, the Novikov ring Λ_ω does arise in the context of Floer homology from a closed 1-form defined on the free loop space of M. This was used by Hofer and Salamon in [32] to prove the Arnold conjecture in the weakly monotone case, and is an indication of the close connection between quantum cohomology and Floer homology.

$$\square$$

Quantum cohomology with Novikov rings

In the monotone case this can be defined much as before. Again, we define

$$QH^*(M) = H^*(M) \otimes \Lambda.$$

where Λ is now is the Novikov ring Λ_ω. It is graded by

$$QH^k(M) = \bigoplus_{j=0}^{2n} H_j(M) \otimes \Lambda_{k-j}.$$

Here we have dropped the notation ω in Λ and denote by Λ_k the elements of the Novikov ring of degree k. Thus a class $a \in QH^k(M)$ may be written in the form

$$a = \sum_A a_A e^{2\pi i A}, \qquad a_A \in H^{k-2c_1(A)}(M). \tag{9.2}$$

The finiteness condition of the Novikov ring now becomes

$$\#\{A \in \Gamma \,|\, a_A \neq 0, \, \omega(A) \leq c\} < \infty$$

for every $c \geq 0$. The Novikov ring Λ_ω acts on $QH^*(M)$ in the obvious way by

$$\lambda * a = \sum_A \sum_B \lambda_{A-B} a_B e^{2\pi i A}.$$

So $QH^*(M)$ is a module over Λ_ω. If the first Chern class vanishes, then $QH^k(M)$ is a module over Λ_ω for every k. Note that in this case the quantum cohomology groups are graded by the integers and we have $QH^k(M) = H^k(M) \otimes \Lambda_\omega$. In general the group $QH^k(M)$ is only a module over the subring $\Lambda_0 \subset \Lambda_\omega$ of all elements of degree zero. In the monotone case, this is just the group ring over the kernel of $\omega : \Gamma \to \mathbb{Z}$. The dimension of $QH^k(M)$ as a module over Λ_0 is

$$\dim_{\Lambda_0} QH^k(M) = \sum_{j \equiv k \,(\mathrm{mod}\ 2N)} b_j$$

where $b_j = \dim H^j(M)$ is the j-th Betti number of M.

Remark 9.2.2 If the first Chern class c_1 does not vanish, then the quantum cohomology is periodic with period $2N$ where N is the minimal Chern number. In other words there is an isomorphism

$$QH^k(M) \cong QH^{k+2N}(M)$$

given by multiplication with a monomial $e^{2\pi i A}$ where A has Chern class $c_1(A) = N$. This isomorphism depends on the choice of the monomial and so is not unique (unless $H_2(M) = \mathbb{Z}$). $\qquad\Box$

Deformed cup product

As before, the deformed cup product is the homomorphism

$$QH^k(M) \times QH^\ell(M) \to QH^{k+\ell}(M)$$

defined by

$$a * b = \sum_A (a * b)_A e^{2\pi i A}$$

for $a \in H^k(M)$ and $b \in H^\ell(M)$, where $(a * b)_A \in H^{k+\ell-2c_1(A)}(M)$ is given by

$$\int_\gamma (a * b)_A = \Phi_A(\alpha, \beta, \gamma) \tag{9.3}$$

for $\gamma \in H_{k+\ell-2c_1(A)}(M)$ where $\alpha = \mathrm{PD}(a) \in H_{2n-k}(M), \beta = \mathrm{PD}(b) \in H_{2n-\ell}(M)$. Here the invariant Φ_A is defined whenever A is a multiple class by the right hand side of the identity in Lemma 9.1.1 (ii). Apart from this, the previous discussion carries over without essential change. For example, formula (8.7) becomes

$$\int_\delta ((a * b) * c)_A = \sum_B \Phi_{A-B}(\xi_B, \gamma, \delta)$$

where $\xi_B = \mathrm{PD}((a * b)_B)$ and $\gamma = \mathrm{PD}(c)$. Thus, in the weakly monotone case, quantum cohomology with coefficients in the Novikov ring Λ_ω is well-defined and has an associative multiplication. Moreover, as before, the constant term in the expansion (9.3) is just the usual triple intersection index and so corresponds to the ordinary cup product $a \cup b$.

Theorem 9.2.3 *Assume that (M, ω) is weakly monotone. Then the deformed cup product on $QH^*(M)$ is associative, distributive, and skew-commutative. Moreover, if $a \in H^0(M)$ or $a \in H^1(M)$ then the deformed cup-product $a * b$ agrees with the ordinary cup-product $a \cup b$. In particular, the canonical generator $1\!\!1 \in H^0(M)$ is the unit element.*

9.3 Calabi-Yau manifolds

There are some interesting phenomena in the structure of the quantum cohomology ring $QH^*(M)$ for general weakly monotone symplectic manifolds which do not appear in the monotone case. For example the homomorphism $\omega : \Gamma \to \mathbb{R}$ need no longer be integral, and this has its consequences for the Novikov ring Λ_ω. The first interesting case for this phenomenon is that of symplectic manifolds with vanishing first Chern class.

A class of symplectic manifolds which satisfy this condition are the **Calabi-Yau manifolds**. These are complex Kähler 3-folds M (6 real dimensions) with $c_1 = 0$. By a theorem of Yau, such manifolds admit Kähler metrics with vanishing Ricci tensor. Calabi-Yau manifolds have found considerable interest in the recent physics literature (see for example [8] and [9]), and the mirror symmetry conjecture which arises in this context has greatly intrigued mathematicians. All this formed part of the motivation for the development of quantum cohomology.

A particular example of a Calabi-Yau manifold is a quintic hypersurface in $\mathbb{C}P^4$. Another such example is a product $M = \mathbb{T}^2 \times X$ where $X \subset \mathbb{C}P^3$ is a quartic hypersurface (the famous $K3$-surface). Recall that in a Calabi-Yau manifold the dimension of the moduli space $\mathcal{M}(A, J)/G$ is zero for every class A and hence, for a generic almost complex structure J, every simple J-holomorphic curve is isolated. It has been conjectured by physicists that, in the case where M is a quintic in $\mathbb{C}P^4$, this should continue to hold for generic complex structures. However, for general Calabi-Yau manifolds with $b_2 > 1$ it is necessary to consider non-integrable almost complex structures in order prove the existence of only finitely many j-holomorphic curves in each given homology class. An explicit counterexample is given by P.M.H. Wilson in [84].

Convergence

An alternative (conjectural) definition of quantum cohomology can be given with $QH^k(M) = H^k(M, \mathbb{C})$ in which the deformed cup product depends on a complex parameter t with sufficiently large real part. For two classes $A \in H^k(M, \mathbb{C})$ and $b \in H^\ell(M, \mathbb{C})$ the product $a *_t b \in H^{k+\ell}(M, \mathbb{C})$ should be defined by

$$\int_\gamma a *_t b = \sum_A \Phi_A(\alpha, \beta, \gamma) e^{-t\omega(A)} \tag{9.4}$$

for $\gamma \in H_{k+\ell}(M, \mathbb{C})$. It was conjectured by physicists that the series should converge for Calabi-Yau manifolds. If this is the case, then the quantum deformation of the cup product is an actual deformation of the ordinary cup product with deformation parameter t, and the ordinary cup product appears as the limit when $t \to \infty$.

Note also that, in the case of Calabi-Yau manifolds, the only interesting cohomology group, from the point of view of quantum cohomology, is $H^{1,1}(M)$ and, because $H^{2,2}(M) \cong H^{1,1}(M)$ we can think of quantum cohomology as a product structure on $H^{1,1}$. In fact all other groups, apart from $H^{0,0}(M) = \mathbb{Z}$ and $H^{3,0}(M) = \mathbb{Z}$ are either zero or, as in the case of $H^{1,2}(M)$, will only have trivial quantum-cup-products. Now the deformation ring of M is a ring structure on $H^{1,2}$ which gives rise to a differential equation similar to the WDVV-equation above. The **mirror symmetry conjecture** states that associated to each Calabi-Yau manifold M there is a mirror manifold M^* such that the rings $H^{1,1}(M^*)$ and $H^{1,2}(M)$ are naturally isomorphic.

Quintic hypersurfaces in $\mathbb{C}P^4$

Consider the hypersurface of degree k in $\mathbb{C}P^4$

$$Z_k = \left\{ [z_0 : \ldots : z_4] \in \mathbb{C}P^4 \mid \sum_{j=0}^{4} z_j{}^k = 0 \right\}.$$

This manifold is simply connected and has Betti numbers

$$b_2 = b_4 = 1, \qquad b_3 = k^4 - 5k^3 + 10k^2 - 10k + 4.$$

In particular the identity $b_2 = b_4 = 1$ follows from the Lefschetz theorem on hyperplane sections. It follows that $\pi_2(Z_k) = \mathbb{Z}$ and the symplectic form ω does not vanish over $\pi_2(Z)$. Moreover the first Chern class of Z_k is given by

$$c_1 = (5 - k)\iota^* h$$

where $h \in H^2(\mathbb{C}P^4, \mathbb{Z})$ is the standard generator of the cohomology of $\mathbb{C}P^4$ and $\iota : Z_k \to \mathbb{C}P^4$ is the natural embedding of Z_k as a hypersurface in $\mathbb{C}P^4$.

Now let $A \in \pi_2(Z_k)$ be the generator of the homotopy group with $\omega(A) > 0$. An explicit representative of A is given, for example, by the holomorphic curve $[z_0 : z_1] \mapsto [z_0 : z_1 : -z_0 : -z_1 : 0]$ when k is odd. Evaluating the first Chern class on this generator gives

$$c_1(A) = 5 - k.$$

So for $k \leq 4$ the manifold Z_k is monotone. For $k \leq 3$ the minimal Chern number is at least 2 and so the quantum cohomology can be defined with the methods of Section 8.1. For $k = 4$ the minimal Chern number is 1. In this case the quantum cohomology can still be defined with the techniques of Section 8.1, but to prove associativity one has to combine the methods of Section 8.2 with those of Section 9.1. For $k > 5$ there are no nonconstant J-holomorphic curves for a generic almost complex structure because either $\omega(A) \leq 0$ or $c_1(A) < 0$. In the latter case, there cannot be any J-holomorphic curves for dimensional reasons because in 6 dimensions $\dim \mathcal{M}(A, J)/G = 2c_1(A) < 0$.

Hence the most interesting case is that of the quintic hypersurface $M = Z_5$ with Chern class zero. This is the archetypal example of a Calabi-Yau manifold. In order to compute the quantum product, we need to calculate all non-zero invariants $\Phi_A(\alpha_1, \alpha_2, \alpha_3)$. We would like the answer to be given in terms of the numbers n_d of simple J-holomorphic curves of degree $d = \omega(A)$. (Here ω is normalized to give an isomorphism $H_2(M) \to \mathbb{Z}$.) The answer should depend only on the original manifold M, and not on some perturbation of J.

For $A \neq 0$ it is easy to check that, for dimensional reasons, the invariant $\Phi_A(\alpha_1, \alpha_2, \alpha_3)$ is non-zero only when each α_i has dimension 4. More precisely, by condition (8.3), the sum of the degrees of the α_i must equal 12. Since the curves in M are all isolated a generic representative of any cycle α_i with degree $\deg(\alpha_i) < 4$ will avoid all nonconstant J-holomorphic curves. Therefore, since $b_4 = 1$, it suffices to calculate

$$\Phi_A(H, H, H)$$

where $H = [Z_5 \cap \mathbb{C}P^3] \in H_4(Z_5)$ is the class of the hyperplane section.

Observe first that, because Z_5 is a hypersurface in $\mathbb{C}P^4$ of degree 5, we have

$$\Phi_0(H, H, H) = H \cdot H \cdot H = [\mathbb{C}P^1] \cdot Z_5 = 5.$$

Moreover, if C is a simple curve in class A of degree d then $C \cdot H = d$. Thus, if we perturb H to general position, C intersects H in d distinct points and it follows easily that the contribution of the curve C to $\Phi_A(H, H, H)$ is exactly d^3.

Now let us consider the contribution of C to $\Phi_{mA}(H, H, H)$. Choose a holomorphic parametrization $u : \mathbb{C}P^1 \to Z_5$ of C, let $\phi : \mathbb{C}P^1 \to \mathbb{C}P^1$ be a rational map of degree m, and consider the graph $\widehat{v} : \mathbb{C}P^1 \to \mathbb{C}P^1 \times Z_5$ of $u \circ \phi$:

$$\widehat{v}(z) = (z, u(\phi(z))).$$

This curve is not regular with respect to the complex structure $i \times J$. It belongs to a family of curves $\mathcal{M}(A_0 + mA, i \times J)$ of dimension $6 + 4m + 2$ while the dimension in the regular case should only be $8 + 2c_1(mA) = 8$. (Here $4m + 2$ is the real dimension of the space Rat_m of parametrizations ϕ. One can also verify that $\widehat{v}$ is non-regular by using Lemma 3.5.1.) Thus, in order to see how C contributes to $\Phi_{mA}(H, H, H)$, one should perturb the complex structure $i \times J$ to a generic element $\widehat{J}$ and count the number of $\widehat{J}$-holomorphic $(A_0 + mA)$-curves near C. By assumption, there are only finitely many simple J-holomorphic mA-curves which are all regular, and these are therefore separated from C. Hence under the perturbation $j \times J \to \widehat{J}$ their graphs move slightly but remain separated from C. Similarly, if $A = m'A'$, the perturbation of any m'-fold cover of an A'-curve is separated from C. Thus one can isolate the contribution from C to $\mathcal{M}(A_0 + mA, \widehat{J})$.

No one has yet performed this calculation. However, Aspinwall and Morrison in [2] give a different but very natural calculation of this contribution using methods coming from algebraic geometry. Their answer is that each m-fold cover of C should also contribute d^3 to $\Phi_{mA}(H, H, H)$. (This means that there should be exactly one element of $\mathcal{M}(A_0 + mA, \widehat{J})$ coming from C.) Thus, their definition gives the following beautiful formula for $a * a$ when $a \in H^2(M)$ is dual to H

$$\int_H a * a = \sum_A \int_H (a * a)_A e^{2\pi i A}$$

$$\begin{aligned}
&= \sum_A \Phi_A(H, H, H) e^{2\pi i A} \\
&= 5 + \sum_{d=1}^{\infty} \sum_{\deg(C)=d} d^3 (q^d + q^{2d} + q^{3d} + \dots) \\
&= 5 + \sum_{d=1}^{\infty} n_d d^3 \frac{q^d}{1 - q^d}.
\end{aligned}$$

Here we use the notation $e^{2\pi i A} = q^d$ where $A \in H_2(M)$ is the unique homology class of degree d. The second sum is over all simple holomorphic curves C of degree d and n_d is the number of such curves.

It seems very likely that the calculation of $a * a$ using the perturbation $\widehat{J}$ will agree with this Aspinwall–Morrison formula. Observe also that, if we insert $q = e^{-t}$ then the deformed cup-product is given by the formula

$$a *_t a = \left(5 + \sum_{d=1}^{\infty} n_d d^3 \frac{e^{-td}}{1 - e^{-td}} \right) a$$

and the convergence problem is now restated in terms of the convergence of the series on the right hand side.

Chapter 10

Floer Homology

There is a completely different approach to quantum cohomology arising from Floer's proof of the Arnold conjecture and the resulting notion of Floer homology (c.f. [18] and [20]). In this chapter, we will actually construct a cohomology theory, rather than a homology theory. We briefly describe the construction of Floer cohomology and outline a proof that it is isomorphic to the (additive) quantum cohomology defined above. There is also a natural ring structure on Floer cohomology, and we discuss the conjecture that this is isomorphic to the deformed cup product in quantum cohomology.

Floer originally developed his homology theory in the context of monotone symplectic manifolds. This was later extended by Hofer and Salamon in [32] and Ono in [60] to the weakly monotone case and it is this extension which we shall explain below.

10.1   Floer's cochain complex

Let (M, ω) be a compact weakly monotone symplectic manifold and let $H_t = H_{t+1} : M \to \mathbb{R}$ be a smooth, 1-periodic family of Hamiltonian functions. Denote by $X_t : M \to TM$ the Hamiltonian vector field defined by $\iota(X_t)\omega = dH_t$ and consider the time dependent Hamiltonian differential equation

$$\dot{x}(t) = X_t(x(t)). \tag{10.1}$$

The solutions of this equation generate a family of symplectomorphisms $\psi_t : M \to M$ via

$$\frac{d}{dt}\psi_t = X_t \circ \psi_t, \qquad \psi_0 = 0.$$

It is not hard to see that the fixed points of the time-1-map $\psi = \psi_1$ are in one-to-one correspondence with the 1-periodic solutions of (10.1). We now explain how to interpret the contractible 1-periodic solutions as the critical points of the symplectic action functional on the universal cover of the space $\mathcal{L}M$ of contractible loops in M.

For every contractible loop $x : \mathbb{R}/\mathbb{Z} \to M$ there exists a smooth map $u : B \to M$ defined on the unit disc $B = \{z \in \mathbb{C} \,|\, |z| \leq 1\}$ which satisfies $u(e^{2\pi i t}) = x(t)$. Two such maps u_1 and u_2 are called **equivalent** if their sum $u_1 \# (-u_2)$ is homologous to zero (in the space $H_2(M)$ of integral homology divided by torsion). We use the notation

$$[x, u_1] \sim [x, u_2]$$

for equivalent pairs and denote by $\widetilde{\mathcal{L}M}$ the space of equivalence classes. The elements of $\widetilde{\mathcal{L}M}$ will also be denoted by $\tilde{x}$. The space $\widetilde{\mathcal{L}M}$ is the unique covering space of $\mathcal{L}M$ whose group of deck transformations is the image $\Gamma \subset H_2(M)$ of the Hurewicz homomorphism $\pi_2(M) \to H_2(M)$. We denote by

$$\Gamma \times \widetilde{\mathcal{L}M} \to \widetilde{\mathcal{L}M} : (A, \tilde{x}) \mapsto A \# \tilde{x}$$

the obvious action of Γ on $\widetilde{\mathcal{L}M}$.

The symplectic action functional $a_H : \widetilde{\mathcal{L}M} \to \mathbb{R}$ is defined by

$$a_H([x, u]) = \int_B u^* \omega + \int_0^1 H_t(x(t)) \, dt$$

and satisfies

$$a_H(A \# \tilde{x}) = a_H(\tilde{x}) + \omega(A). \tag{10.2}$$

This function can therefore be interpreted as a closed 1-form on the loop space $\mathcal{L}M$ rather than a function on the covering space $\widetilde{\mathcal{L}M}$, which is exactly the situation considered by Novikov: cf Remark 9.2.1.

It is not hard to check that the critical points of a_H are precisely the equivalence classes $[x, u]$ where $x(t) = x(t+1)$ is a contractible periodic solution of (10.1). We shall denote by $\widetilde{\mathcal{P}}(H) \subset \widetilde{\mathcal{L}M}$ the set of critical points and by $\mathcal{P}(H) \subset \mathcal{L}M$ the corresponding set of periodic solutions. Floer homology is essentially an infinite dimensional version of Morse-Novikov theory for the symplectic action functional.

Consider the (upwards) gradient flow lines of a_H with repect to an L^2-metric on $\mathcal{L}M$ which is induced by an almost complex structure on M. These are solutions $u : \mathbb{R}^2 \to M$ of the partial differential equation

$$\frac{\partial u}{\partial s} + J(u)\frac{\partial u}{\partial t} - \nabla H_t(u) = 0 \tag{10.3}$$

with periodicity condition $u(s, t+1) = u(s, t)$ and limit condition

$$\lim_{s \to \pm\infty} u(s, t) = x^{\pm}(t) \tag{10.4}$$

where $x^{\pm} \in \mathcal{P}(H)$. We denote by $\mathcal{M}(\tilde{x}^-, \tilde{x}^+) = \mathcal{M}(\tilde{x}^-, \tilde{x}^+, H, J)$ the space of all solutions of (10.3) and (10.4) with $\tilde{x}^- \# u = \tilde{x}^+$. The elements of this space have finite energy

$$E(u) = \tfrac{1}{2} \int_0^1 \int_{-\infty}^{\infty} \left(\left|\frac{\partial u}{\partial s}\right|^2 + \left|\frac{\partial u}{\partial t} - X_t(u)\right|^2 \right) \, ds\, dt = a_H(\tilde{x}^+) - a_H(\tilde{x}^-).$$

Moreover, for a generic Hamiltonian function $H : M \times \mathbb{R}/\mathbb{Z} \to \mathbb{R}$, the space $\mathcal{M}(\tilde{x}^-, \tilde{x}^+)$ is a finite dimensional manifold of dimension

$$\dim \mathcal{M}(\tilde{x}^-, \tilde{x}^+) = \mu(\tilde{x}^+) - \mu(\tilde{x}^-).$$

Here the function $\mu : \widetilde{\mathcal{P}}(H) \to \mathbb{Z}$ is a version of the Maslov index due to Conley and Zehnder. The integer $\mu([x,u])$ is defined by trivializing the tangent bundle over the disc $u(B)$ and considering the path of symplectic matrices generated by the linearized Hamiltonian flow along $x(t)$. We refer to [75] and [15] for more details. Here we only point out that

$$\mu(A\#\tilde{x}) = \mu(\tilde{x}) + 2c_1(A) \tag{10.5}$$

for $\tilde{x} \in \widetilde{\mathcal{P}}(H)$ and $A \in \Gamma$. Moreover, the index can be normalized so that

$$\mu(\tilde{x}) = \mathrm{ind}_H(x) \tag{10.6}$$

whenever $H_t \equiv H$ is a C^2-small Morse function and $\tilde{x} = [x,u]$, where $x(t) \equiv x$ is a critical point of H and $u(z) = x$ is the constant disc. The right hand side in (10.6) is then to be understood as the Morse index.

In the case $\mu(\tilde{y}) - \mu(\tilde{x}) = 1$, the space $\mathcal{M}(\tilde{x},\tilde{y})$ is a 1-dimensional manifold, and so each point in the quotient $\mathcal{M}(\tilde{x},\tilde{y})/\mathbb{R}$ (with $\mathbb{R}$ acting by time shift) is isolated. Moreover, for a generic Hamiltonan H we have the following finiteness result.

Proposition 10.1.1 *For a generic almost complex structure J and generic Hamiltonian H we have*

$$\sum_{\substack{\omega(A) \leq c \\ c_1(A) = 0}} \# \left\{ \mathcal{M}(\tilde{x}, A\#\tilde{y})/\mathbb{R} \right\} < \infty$$

for all $\tilde{x}, \tilde{y} \in \widetilde{\mathcal{P}}(H)$ with $\mu(\tilde{y}) - \mu(\tilde{x}) = 1$ and every constant c.

To prove this, one has to show that the relevant moduli spaces are compact. This will be the case if no bubbling occurs. The key observation is that, for a generic almost complex structure J, the set of points lying on a J-holomorphic sphere of Chern number 0 forms a set in M of codimension 4 and so, for a generic H, no such sphere will intersect an isolated connecting orbit. Thus, it follows from Gromov's compactness theorem that they cannot bubble off. Moreover, J-holomorphic spheres of negative Chern number do not exist by weak monotonicity. J-holomorphic spheres of Chern number at least 1 connot bubble off because otherwise in the limit there would be a connecting orbit with negative index difference but such orbits do not exist generically. This is the essence of the proof of Proposition 10.1.1. Details are carried out in [32].

Whenever $\mu(\tilde{y}) - \mu(\tilde{x}) = 1$ we denote

$$n(\tilde{x},\tilde{y}) = \# \left\{ \mathcal{M}(\tilde{x},\tilde{y})/\mathbb{R} \right\},$$

where the connecting orbits are to be counted with appropriate signs determined by a system of coherent orientations of the moduli spaces of connecting orbits as in [21]. These numbers determine a cochain complex as follows. Define

$$CF^k = CF^k(H)$$

as the set of formal sums

$$\xi = \sum_{\mu(\tilde{x})=k} \xi_{\tilde{x}} \langle \tilde{x} \rangle$$

which satisfy the finiteness condition

$$\# \left\{ \tilde{x} \in \widetilde{\mathcal{P}}(H) \mid \xi_{\tilde{x}} \neq 0,\ a_H(\tilde{x}) \leq c \right\} < \infty.$$

Here the generators $\langle \tilde{x} \rangle$ of the Floer complex run over the set $\widetilde{\mathcal{P}}(H)$ of critical points of the action functional, and the coefficients $\xi_{\tilde{x}}$ may be taken in $\mathbb{Z}$ or in $\mathbb{Q}$. This complex CF^* is a module over the Novikov ring $\Lambda = \Lambda_\omega$ defined above, with action given by the formula

$$\lambda * \xi = \sum_{\tilde{x}} \sum_A \lambda_A \xi_{(-A)\#\tilde{x}} \ \langle \tilde{x} \rangle.$$

Note that this action will change the degree, unless λ_A is nonzero only when $c_1(A) = 0$. Further, the dimension of $CF^*(H)$ as a module over Λ_ω is precisely the number $\#\mathcal{P}(H)$ of contractible periodic solutions of the Hamiltonian system (10.1).

The above numbers $n(\tilde{x}, \tilde{y})$ determine a coboundary operator

$$\delta : CF^*(H) \to CF^*(H)$$

defined by

$$\delta\tilde{x} = \sum_{\mu(\tilde{y})=k+1} n(\tilde{x}, \tilde{y})\langle \tilde{y} \rangle.$$

Proposition 10.1.1 and the formula (10.2) guarantee the finiteness condition required for

$$\delta\tilde{x} \in CF^{k+1}(H).$$

Floer's proof that the square of this operator is zero carries over to the weakly monotone case. Here the key observation is that 1-parameter families of connecting orbits with index difference 2 will still avoid the J-holomorphic spheres of Chern number 0 because they form a 3-dimensional set in M while these J-holomorphic spheres form a set of codimension 4. Similarly, holomorphic spheres of Chern number 1 can only bubble off if they intersect a periodic solution, and this does not happen for a generic H because the points on these spheres form a set in M of codimension 2 while the periodic orbits form 1 dimensional sets. For J-holomorphic spheres with Chern number at least 2 the same argument as above applies. Hence no bubbling occurs for connecting orbits with index difference 2 and hence such orbits can only degenerate by splitting into a pair of orbits each with index difference 1. As in the standard theory (cf. [20], [43], [75]) this shows that $\delta \circ \delta = 0$. Hence the solutions of (10.3) determine a cochain complex (CF^*, δ), and its homology groups

$$HF^*(M, \omega, H, J) = \frac{\ker \delta}{\operatorname{im} \delta}$$

are called the **Floer cohomology groups** of the pair (H, J). Because the coboundary map is linear over Λ_ω it follows that the Floer cohomology groups form a module over Λ_ω. In [32] it is proved that the Floer cohomology groups are independent of the almost complex structure J and the Hamiltonian H used to define them.

Theorem 10.1.2 *For two pairs (H^α, J^α) and (H^β, J^β), which satisfy the regularity requirements for the definition of Floer cohomology, there exists a natural isomorphism*

$$\Phi^{\beta\alpha} : HF^*(M, \omega, H^\alpha, J^\alpha) \to HF^*(M, \omega, H^\beta, J^\beta).$$

If (H^γ, J^γ) is another such pair then

$$\Phi^{\gamma\alpha} = \Phi^{\gamma\beta} \circ \Phi^{\beta\alpha}, \qquad \Phi^{\alpha\alpha} = \mathrm{id}.$$

These isomorphisms are linear over Λ_ω.

The theorem is proved by choosing a homotopy (H_s, J_s) from (H^α, J^α) to (H^β, J^β) and considering the finite energy solutions of the following time dependent version of equation (10.3)

$$\frac{\partial u}{\partial s} + J_s(u)\frac{\partial u}{\partial t} - \nabla H_{s,t}(u) = 0.$$

Here u satisfies the usual periodicity condition $u(s, t+1) = u(s,t)$. Any such solution will have limits

$$\lim_{s \to -\infty} u(s,t) = x^\alpha(t), \qquad \lim_{s \to +\infty} u(s,t) = x^\beta(t)$$

where $x^\alpha \in \mathcal{P}(H^\alpha)$ and $x^\beta \in \mathcal{P}(H^\beta)$. These solutions determine a chain map $CF^*(H^\alpha) \to CF^*(H^\beta)$ which is of degree zero and, choosing a homotopy of homotopies, one can see that the induced map on Floer cohomology is independent of the choice of the homotopy. These arguments are again precisely the same as in Floer's original proof in [20] for the monotone case, and for the present case they are carried out in [32]. (See also [75] for the case $c_1 = [\omega] = 0$.)

Now one can specialize to a time independent Hamiltonian function and prove that the Floer cohomology groups are naturally isomorphic to the cohomology of the underlying manifold M with coefficients in the Novikov ring Λ_ω. But these are precisely the quantum cohomology groups of M. (In order not to be concerned with torsion, we take here cohomology with rational coefficients, replacing Λ_ω by $\Lambda_\omega \otimes \mathbb{Q}$.)

The following theorem was proved by Floer [20] in the monotone case and by Hofer and Salamon [32] in the case where either $c_1(A) = 0$ for all $A \in \pi_2(M)$ or the minimal Chern number is $N \geq n$. The general case is treated with different methods in [63].

Theorem 10.1.3 *Assume that (M, ω) is weakly monotone. Then, for every regular pair (H^α, J^α), there exists an isomorphism*

$$\Phi^\alpha : HF^*(M, \omega, H^\alpha, J^\alpha) \to QH^*(M),$$

where the cohomology groups are to be understood with rational coefficients. These maps are natural in the sense that

$$\Phi^\beta \circ \Phi^{\beta\alpha} = \Phi^\alpha$$

and they are linear over Λ_ω.

Floer's original proof for the monotone case and the proof in [32] are based on the following idea. Choose a time independent Hamiltonian $H : M \to \mathbb{R}$ which is a Morse function. If H is sufficiently small in the C^2-norm then the 1-periodic solutions of (10.1) are precisely the critical points of H and, by (10.6), their Maslov index agrees with the Morse index. Moreover, the gradient flow lines $u : \mathbb{R} \to M$ of H are solutions of the ordinary differential equation

$$\dot{u}(s) = \nabla H(u(s))$$

and they form special solutions of the partial differential equation (10.3), namely those which are independent of t. These solutions determine the Morse-Witten coboundary operator

$$\delta_{\mathrm{MW}} : C^*(H) \to C^*(H).$$

This coboundary operator is defined on the same cochain complex as the Floer coboundary δ and, by (10.6), the cochain complex has the same grading for both theories. But the homology of the Morse-Witten coboundary operator is naturally isomorphic to the quantum cohomology of M

$$QH^*(M) = \frac{\ker \delta_{\mathrm{MW}}}{\operatorname{im} \delta_{\mathrm{MW}}}.$$

(See [86], [73], or Schwarz's book [76].) Thus, to prove Theorem 10.1.3 one must show that all the solutions of (10.3) and (10.4) with

$$\mu(u) = \mu(\tilde{x}^+) - \mu(\tilde{x}^-) \leq 1$$

are independent of t, provided that $H_t \equiv H$ is a Morse function which is independent of t and is rescaled by a sufficiently small factor.

To prove this result without any bound on the energy of the solution requires one to assume that either M is monotone, or $c_1(A) = 0$ for all $A \in \pi_2(M)$, or the minimal Chern number is $N \geq n$. In the other cases of weak monotonicity (where the minimal Chern number is $N = n - 1$ or $N = n - 2$) it has so far only been possible to prove t-independence for the solutions of (10.3) with a given bound on the energy, with the required smallness of H depending on this bound. To use such a result for the proof of Theorem 10.1.3 one needs an alternative definition of Floer cohomology. (First truncate the chain complex and then take inverse and direct limits.) These ideas are due to Ono and in [60] he proved Theorem 10.1.3 with this modified definition of Floer cohomology.

To prove this theorem with the original definition of the Floer groups requires a different approach which was found by Piunikhin–Salamon–Schwarz [63]. The idea is to consider perturbed J-holomorphic planes $u : \mathbb{C} \to M$ such that $u(re^{2\pi it})$ converges to a periodic solution $x^\alpha(t)$ of the (time dependent) Hamiltonian system H^α as $r \to \infty$ and such that $u(0)$ lies on the unstable manifold of a given critical point x of the Morse function $H : M \to \mathbb{R}$ (with respect to the upward gradient flow). One can think of these as J-holomorphic **spiked disks** where the spike is the gradient flowline from x to $u(0)$. In the case where the index difference is zero the moduli space of such spiked disks is 0-dimensional and hence the numbers $n(x, \tilde{x}^\alpha)$ of its elements can be used to construct a chain map $C^*(H) \to CF^*(H^\alpha)$. In [63] it is shown that this map induces an isomorphism on cohomology.

As a corollary of Theorem 10.1.3 it follows that all weakly monotone symplectic manifolds (M, ω) satisfy the **Arnold conjecture**. To see this recall that the dimension of the Floer chain complex $CF^*(H^\alpha)$ as a module over the Novikov ring is the number of contractible periodic solutions.

Corollary 10.1.4 (Arnold conjecture) [20, 32, 60] *Let (M, ω) be a weakly monotone compact symplectic manifold and $\psi : M \to M$ be a Hamiltonian symplectomorphism with only nondegenerate fixed points. Then*

$$\#\mathrm{Fix}(\psi) \geq \sum_{j=0}^{2n} b_j(M)$$

where $b_j(M) = \dim H_j(M)$ denote the Betti numbers of M.

10.2 Ring structure

There is a natural ring structure on Floer cohomology which is defined using trajectories which connect three periodic orbits, in contrast to the cylindrical trajectories used above to define the boundary operator. The domain of these trajectories is the "pair of pants" Σ, and we begin by describing this space.

Σ is a noncompact Riemann surface of genus zero with three cylindrical ends, provided with a conformal structure which is the standard product structure on the cylindrical ends. To define this precisely, fix parametrizations

$$\phi_1, \phi_2 : (-\infty, 0) \times \mathbb{R}/\mathbb{Z} \to \Sigma, \qquad \phi_3 : (0, \infty) \times \mathbb{R}/\mathbb{Z} \to \Sigma$$

of the cylindrical ends with disjoint images $U_j = \phi_j((-\infty, 0) \times \mathbb{R}/\mathbb{Z})$ for $j = 1, 2$ and $U_3 = \phi_3((0, \infty) \times \mathbb{R}/\mathbb{Z})$. Assume that the complement

$$\Sigma' = \Sigma - U_1 - U_2 - U_3$$

is diffeomorphic to a 2-sphere with three open discs removed. In particular Σ' is compact. Choose a complex structure on Σ which on the three cylindrical ends pulls back to the standard structure $s + it$.

In order to define the appropriate trajectory space, choose three Hamiltonian functions $H_j : \mathbb{R} \times \mathbb{R}/\mathbb{Z} \times M \to \mathbb{R}$ which vanish for $|s| \leq 1/2$ and are independent of s for $|s| \geq 1$. Then consider the space of smooth maps

$$u : \Sigma \to M$$

such that the restriction of u to Σ' is a J-holomorphic curve and $u_j(s, t) = u \circ \phi_j(s, t)$ satisfies

$$\frac{\partial u_j}{\partial s} + J(u_j)\frac{\partial u_j}{\partial t} - \nabla H_j(s, t, u_j) = 0$$

for $j = 1, 2, 3$. Here the gradient in the last term is to be understood with respect to the third argument which lies in M. Any finite energy solution of these equations satisfies the limit conditions

$$\lim_{s \to -\infty} u_1(s, t) = x_1(t), \quad \lim_{s \to -\infty} u_2(s, t) = x_2(t), \quad \lim_{s \to +\infty} u_3(s, t) = x_3(t)$$

where $x_j \in \mathcal{P}(H_j)$. (Here we slightly abuse notation and denote by $\mathcal{P}(H_j)$ the periodic solutions of the Hamiltonian $H_j(-\infty, t, x)$ for $j = 1, 2$ and of $H_j(+\infty, t, x)$

for $j = 3$.) The space of such solutions u in the correct homology class will be denoted by $\mathcal{M}(\tilde{x}_1, \tilde{x}_2, \tilde{x}_3)$. It has dimension

$$\dim \mathcal{M}(\tilde{x}_1, \tilde{x}_2, \tilde{x}_3) = \mu(\tilde{x}_3, H_3) - \mu(\tilde{x}_1, H_1) - \mu(\tilde{x}_2, H_2)$$

In the zero dimensional case we get finitely many solutions by the same argument as above.

By the usual counting procedure, these finitely many solutions determine a chain map

$$CF^k(H_1) \times CF^\ell(H_2) \to CF^{k+\ell}(H_3) : (\xi, \eta) \mapsto \xi * \eta.$$

It follows by the usual gluing techniques in Floer homology that this map is a cochain homomorphism and therefore induces a homomorphism of Floer cohomologies

$$HF^k(H_1) \times HF^\ell(H_2) \to HF^{k+\ell}(H_3).$$

In view of its construction this map is called the **pair-of-pants** product. It is easy to see, by the usual deformation arguments, that the pair-of-pants product is independent of the conformal structure on the Riemann surface Σ used to define it. It is also independent of the Hamiltonian functions H_j, as long as they are not changed at $\pm\infty$. If they are changed at $\pm\infty$, one again uses Floer's gluing techniques to prove that the product is natural with respect to the isomorphisms $\Phi^{\beta\alpha}$ of Theorem 10.1.2. To prove that the product is skew-commutative with the usual sign conventions, choose an orientation preserving diffeomorphism $f : \Sigma \to \Sigma$ which interchanges the two cylindrical ends on the left. This diffeomorphism changes the conformal structure of Σ but, as we have seen above, the resulting map on Floer cohomology is independent of the choice of the conformal structure on Σ. The proof of associativity requires the gluing of two surfaces $\Sigma_{12,3}$ and $\Sigma_{34,5}$ with 3 cylindrical ends over a very long cylinder. This process cancels the two ends labelled by 3 and results in a Riemann surface $\Sigma_{124,5}$ with 4 cylindrical ends. If the *neck* is sufficiently long then the resulting *triple product* corresponds to the composition $(\xi * \eta) * \zeta$. Now vary the conformal structure on the surface $\Sigma_{124,5}$ and decompose it in a different way to obtain the identity $(\xi * \eta) * \zeta = \xi * (\eta * \zeta)$. This variation of the conformal structure corresponds to the moving of the point z in the proof of the associativity of quantum cohomology in Section 8.2.

An alternative definition of a product structure on Floer cohomology which uses differential forms is presented by Viterbo in [83].

10.3 A comparison theorem

According to Theorem 10.1.3 the Floer cohomology groups $HF^*(M)$ are naturally isomorphic to the quantum cohomology groups $QH^*(M)$ and one would expect the pair-of-pants product to correspond to the quantum deformation of the cup product under this isomorphism. This question is already nontrivial in the case $\pi_2(M) = 0$ where the quantum deformation agrees with the ordinary cup product. This was recently proved by Schwarz [77] in his thesis for the case $\pi_2(M) = 0$, by Piunikhin–Salamon–Schwarz [63] for the general case, and independently by Liu [39] and Ruan–Tian [69].

Theorem 10.3.1 *Let $\Phi^\alpha : HF^*(M,\omega,H^\alpha,J^\alpha) \to QH^*(M)$ be the isomorphisms of Theorem 10.1.3 and similarly for Φ^β and Φ^γ. Then*

$$\Phi^\gamma(\xi^\alpha * \xi^\beta) = \Phi^\alpha(\xi^\alpha) * \Phi^\beta(\xi^\beta)$$

for $\xi^\alpha \in HF^*(H^\alpha)$ and $\xi^\beta \in HF^*(H^\beta)$. Here the product on the left is defined by the pair-of-pants construction while the product on the right is the quantum deformation of the cup product.

The proof of this theorem goes along the following lines. First use the Morse-Witten complex to represent the quantum deformation of the cup product in a somewhat different way. Given three Morse functions $H_1, H_2, H_3 : M \to \mathbb{R}$, three critical points x_1, x_2, x_3, and three distinct points $z_1, z_2, z_3 \in \mathbb{C}P^1$, consider the space

$$\mathcal{M}_A(x_1, x_2, x_3) = \mathcal{M}_A(x_1, x_2, x_3; H, \mathbf{z})$$

of all J-holomorphic A-spheres $u : \mathbb{C}P^1 \to M$ such that

$$u(z_1) \in W^u(x_1, H_1), \qquad u(z_2) \in W^u(x_2, H_2), \qquad u(z_3) \in W^s(x_3, H_3).$$

Here $W^s(x, H)$ and $W^u(x, H)$ denote the stable and unstable manifolds of a critical point x of H with respect to the (upward) gradient flow of the Morse function H. One can think of such objects as **spiked J-holomorphic spheres**. Since

$$\dim W^s(x, H) = \mathrm{ind}_H(x), \qquad \dim W^u(x, H) = 2n - \mathrm{ind}_H(x),$$

we have generically

$$\dim \mathcal{M}_A(x_1, x_2, x_3) = 2c_1(A) + \mathrm{ind}_{H_3}(x_3) - \mathrm{ind}_{H_1}(x_1) - \mathrm{ind}_{H_2}(x_2).$$

Whenever this dimension is zero, denote by $n_A(x_1, x_2, x_3)$ the number of points in $\mathcal{M}_A(x_1, x_2, x_3)$ counted with appropriate signs. This gives rise to a chain map

$$C^*(H_1) \times C^*(H_2) \to C^*(H_3)$$

defined by

$$\langle x_1 \rangle * \langle x_2 \rangle = \sum_A n_A(x_1, x_2, x_3)\langle x_3 \rangle, \tag{10.7}$$

and this induces a map on quantum cohomology. Here the boundary map is taken to be δ_{MW} and is determined by counting the connecting orbits of the finite dimensional gradient flows. This product agrees with the deformed cup product because the stable and unstable manifolds of critical points of Morse functions represent cohomology classes which generate the cohomology of M.

Now it follows from the standard gluing arguments in Floer homology that the isomorphism $\Phi^\alpha : HF^*(H^\alpha) \to QH^*(M)$ of Theorem 10.1.3, as described in [63], intertwines the two product structures. To see this one has to glue three J-holomorphic spiked disks to the *boundary* of a J-holomorphic pair-of-pants. More precisely, the boundary components are cylindrical ends abutting on periodic solutions and one uses Floer's gluing theorem. As a result one obtains a (perturbed) J-holomorphic sphere with three spikes as described above. More details of this argument are given in [63]. The full details of the proof will appear in Schwarz [78]. A physicist's approach to this problem may be found in Sadov [72]. Note also that Theorem 10.3.1 gives rise to an alternative proof of the associativity of quantum cohomology.

10.4   Donaldson's quantum category

Symplectomorphisms

The Floer homology approach to quantum cohomology can be extended in two
directions. Assume for simplicity that the symplectic manifold (M, ω) is compact,
simply connected, and monotone. Then there are Floer cohomology groups $HF^*(\phi)$
for every symplectomorphism ϕ. As above they are graded modulo $2N$ where N is
the minimal Chern number, and the Euler characteristic of the theory

$$\chi(HF^*(\phi)) = L(\phi)$$

is the Lefschetz number of ϕ. The critical points that generate the Floer complex are now the fixed points of ϕ, which we assume to be all nondegenerate, and
the connecting orbits are J-holomorphic maps $u : \mathbb{R} \times [0,1] \to M$ which satisfy
$u(s,1) = \phi(u(s,0))$. We may think of these as J-holomorphic sections of a symplectic fiber bundle $P \to \mathbb{R} \times S^1$ with fiber M and holonomy ϕ around S^1. If there
are degenerate fixed points then we must choose a Hamiltonian perturbation as in
the previous section. The resulting Floer homology groups are independent of the
Hamiltonian perturbation and the almost complex structure used to define them
and they only depend on the symplectic isotopy class of ϕ. (See [20], [22], and [14]
for more details.) Now for any two symplectomorphisms ϕ and ψ there is a natural
isomorphism

$$HF^*(\phi) \to HF^*(\psi \circ \phi \circ \psi^{-1}).$$

Moreover, according to Donaldson, there is an analogue of the deformed cup-
product, namely a symmetric and associative pairing

$$HF^*(\phi) \otimes HF^*(\psi) \to HF^*(\psi \circ \phi).$$

This should be defined in terms of J-holomorphic sections of a symplectic fiber
bundle $P \to S$ with fiber M where S is a 2-sphere with three punctures and the
holonomies around these punctures are conjugate to ϕ, ψ, and $\psi \circ \phi$, respectively.
This product structure can be interpreted as a **category** in which the objects are
the symplectomorphisms of M and and the morphisms from ϕ to ψ are the elements
of the Floer cohomology group $HF^*(\psi \circ \phi^{-1})$. Composition of two morphisms is
given by the quantum product.

Now in the case $\psi = \mathrm{id}$ it follows from the discussion of Section 10.1 that, in
the monotone case, the Floer cohomology groups are isomorphic to the ordinary
cohomology groups $HF^*(\mathrm{id}) = QH^*(M) = H^*(M)$, where the grading is made
periodic with period $2N$. Thus the cohomology of M acts on the Floer cohomology
of ϕ

$$H^*(M) \otimes HF^*(\phi) \to HF^*(\phi).$$

In this case there is yet another way to think of the cuantum product structure.
Namely, one can represent a k-dimensional cohomology class by a codimension-k
submanifold (or pseudocycle) $V \subset M$ and then intersect the spaces of connecting
orbits in the construction of the Floer homology groups of ϕ with this submanifold.
More generally, one can intersect these connecting orbit spaces with finite dimen-
sional submanifolds of the path space $\Omega_\phi = \{\gamma : [0,1] \to M \mid \gamma(1) = \phi(\gamma(0))\}$ and
obtain an action of the (low dimensional) cohomology $H^*(\Omega_\phi)$ on $HF^*(\phi)$.

Specializing further to the case $\phi = \text{id}$, we see that the above product construction agrees with the *pair-of-pants product* of Section 10.2, and so in this case the Donaldson category reduces to the quantum cohomology of M.

Mapping tori

The previous structures are particularly interesting when the symplectic manifold M is the moduli space M_Σ of flat connections over a Riemann surface Σ (of large genus) with structure group SU(2) or SO(3). In this case the mapping class group of Σ acts on the manifold M_Σ by symplectomorphisms $\phi_f : M_\Sigma \rightarrow M_\Sigma$ (where $f : \Sigma \rightarrow \Sigma$ is an orientation preserving diffeomorphism) and one can examine the Floer cohomology groups $HF^*(\phi_f)$ generated by the mapping class group. Results in this direction will appear in the thesis of M. Callaghan [7]. Now there is an alternative construction, based on Floer cohomology groups for 3-manifolds and Yang-Mills instantons on 4-dimensional cobordisms. Every diffeomorphism $f : \Sigma \rightarrow \Sigma$ induces a mapping torus $Y_f = \Sigma \times \mathbb{R}/ \sim$ where $\Sigma \times \{t\}$ is identified with $\Sigma \times \{t + 1\}$ via f. These 3-manifolds determine Floer homology groups $HF^*(Y_f)$ constructed from flat SO(3)-connections on Y_f and anti-self-dual instantons on $Y_f \times \mathbb{R}$. It was conjectured by Atiyah and Floer, and proved in Dostoglou–Salamon [16], that there is a natural isomorphism

$$HF^*(\phi_f) \cong HF^*(Y_f).$$

Now in Floer-Donaldson theory there is a pairing

$$HF^*(Y_f) \otimes HF^*(Y_g) \rightarrow HF^*(Y_{gf})$$

determined by anti-self-dual instantons over the 4-manifold X which is fibered over the 3-punctured sphere S with fiber S and holonomy f, g, and gf. Of course, it is natural to conjecture that the two product structures should be preserved by the above isomorphisms and this will be proved in Salamon [74].

An interesting special case arises when $g = \text{id}$. In this case the symplectic Floer cohomology of ϕ_g and hence the instanton Floer cohomology of Y_g agrees with the ordinary cohomology of the moduli space M_Σ made periodic with period 4. So $H^*(M_\Sigma)$ acts on the Floer cohomology of Y_f

$$H^*(M_\Sigma) \otimes HF^*(Y_f) \rightarrow HF^*(Y_f).$$

Now the cohomology of M_Σ is well understood and is closely related to the homology of the Riemann surface Σ itself. For example there is a universal construction of a homomorphism $\mu : H_1(\Sigma) \rightarrow H^3(\mathcal{B}_\Sigma)$ where $\mathcal{B}_\Sigma = \mathcal{A}_\Sigma/\mathcal{G}_\Sigma$ denotes the infinite dimensional configuration space of connections on the bundle $P \rightarrow \Sigma$ modulo gauge equivalence. This is Donaldson's μ-map. It can be roughly described as the slant-product

$$\mu(\gamma) = -\frac{1}{4}p_1(\mathcal{P})/\gamma$$

where $p_1(\mathcal{P}) \in H^4(\mathcal{B}_\Sigma \times \Sigma)$ denotes the first Pontryagin class of the universal SO(3)-bundle $\mathcal{P} \rightarrow \mathcal{B}_\Sigma \times \Sigma$. In more explicit terms, the induced cohomology class in $H^3(M_\Sigma)$ is represented by the codimension-3-submanifold $V_\gamma \subset M_\Sigma$ of those flat connections which have trivial holonomy around γ. In summary, we have the following diagrams

$$
\begin{array}{ccc}
HF^*(Y_f) & \overset{\mu(\gamma)}{\to} & HF^*(Y_f) \\
\downarrow & & \downarrow \\
HF^*(\phi_f) & \overset{V_\gamma}{\to} & HF^*(\phi_f)
\end{array}
\quad , \quad
\begin{array}{ccc}
HF^*(Y_f) & \overset{X_\gamma}{\to} & HF^*(Y_f) \\
\downarrow & & \downarrow \\
HF^*(\phi_f) & \to & HF^*(\phi_f)
\end{array}
\quad .
$$

In each diagram the vertical arrows are the natural isomorphisms of [16]. In the diagram on the left the horizontal maps are defined by cutting down the moduli spaces of connecting orbits by intersecting them with suitable submanifolds. In the diagram on the right the horizontal maps are given by the product structures which are defined in terms of cobordisms. For example, the class $\gamma \in H_1(\Sigma)$ determines a natural cobordism X_γ with boundary $\partial X_\gamma = (-Y_f) \cup Y_f$. Of course, all four definitions of the product should agree under the natural isomorphisms. The relations between these product structures play an important role in the work of Callaghan about symplectic isotopy problems on M_Σ and a detailed discussion will appear in his thesis [7].

These product structures also play an important role in the Floer-Fukaya construction of cohomology groups $HFF^*(Y,\gamma)$ associated to pairs (Y,γ) where Y is a 3-manifold and $\gamma \in H_1(Y)$. In another direction, the homomorphisms in Floer's exact sequence can be interpreted in terms of these product structures and any analogue in symplectic Floer theory should be related to quantum cohomology.

Lagrangian intersections

There are similar structures in Floer cohomology for Lagrangian intersections. These form in fact the original context of Donaldson's quantum category construction. With M as above (compact, simply connected, and monotone) there are Floer cohomology groups $HF^*(L_0, L_1)$ for every pair of Lagrangian submanifolds L_0 and L_1 with $H^1(L_i, \mathbb{R}) = 0$. In this case the critical points are the intersection points $L_0 \cap L_1$ and the connecting orbits are J-holomorphic curves $u : \mathbb{R} \times [0,1] \to M$ with $u(s,0) \in L_0$ and $u(s,1) \in L_1$. This is the context of Floer's original work in [18]. The Euler charcteristic of Floer cohomology is now the intersection number

$$
\chi(HF^*(L_0, L_1)) = L_0 \cdot L_1.
$$

Again there is a pairing

$$
HF^*(L_0, L_1) \otimes HF^*(L_1, L_2) \to HF^*(L_0, L_2)
$$

defined by holomorphic triangles. In [18] Floer proved that $HF^*(L_0, L_0)$ is isomorphic to the ordinary cohomology of L_0 provided that $\pi_2(M) = 0$. Under our assumptions, where M is simply connnected, this condition is never satisfied and the corresponding assertion is an open question.

This multiplicative structure can be interpreted as a quantum category where the objects are the Lagrangian submanifolds $L \subset M$ (with $H^1(L, \mathbb{R}) = 0$) and the morphisms from L_0 to L_1 are the elements of the Floer cohomology group $HF^*(L_0, L_1)$. The above structur with symplectomorphisms is a special case of this with $M = N \times N$, $L_0 = \Delta$, $L_1 = \operatorname{graph}(\phi)$, and $L_2 = \operatorname{graph}(\psi)$.

Heegard splittings

The Floer theory for Lagrangian intersections is related to 3-manifolds as follows. If Y is a homology-3-sphere choose a Heegard splitting $Y = Y_0 \cup Y_1$ over a Riemann surface Σ. Then each handlebody Y_j determines a Lagrangian submanifold $L_j = L_{Y_j} \subset M_\Sigma$ and, according to Atiyah [4] and Floer, there should be a natural isomorphism

$$HF^*(Y_0 \cup Y_1) \cong HF^*(L_0, L_1).$$

As before, there is a product structure

$$HF^*(Y_0 \cup Y_1) \otimes HF^*(Y_1, Y_2) \to HF^*(Y_0, Y_2)$$

defined directly in terms of Yang-Mills instantons on a suitable 4-dimensional cobordism and the two product structures should be related by the above (conjectural) isomorphisms.

10.5 Closing remark

We close this book with the observation that both in the definition of Floer cohomology and in the definition of quantum cohomology the same difficulties arise from the presence of J-holomorphic spheres with negative Chern number. In order to extend either theory to general compact symplectic manifolds one has to find techniques of dealing with J-holomorphic curves of negative Chern number and their multiple covers. For example an important putative theorem would be that multiply covered J-holomorphic curves of negative Chern number cannot be approximated by simple curves (in the same homology class). In the almost complex case no one has so far developed such techniques. One would expect that if such methods can be found then they should give rise to both an extension of quantum cohomology to general compact symplectic manifolds and a proof of the Arnold conjecture in the general case.

Appendix A

Gluing

In this section we give a proof of the decomposition formula in Lemma 8.2.5. This proof is based on a gluing theorem for J-holomorphic curves which is the converse of Gromov's compactness theorem. It asserts, roughly speaking, that if two (or more) J-holomorphic curves intersect and satisfy a suitable transversality condition then they can be approximated, in the *weak sense* of Section 4.4, by a sequence of J-holomorphic curves representing the sum of their homology classes. The proof is along similar lines as the Taubes gluing theorem for anti-self-dual instantons on 4-manifolds. Our proof is an adaption of the argument in Donaldson–Kronheimer [13], page 287–295, to the two dimensional case of J-holomorphic curves.

We shall consider two J-holomorphic curves $u \in \mathcal{M}(A, J)$ and $v \in \mathcal{M}(A, J)$ such that $u(0) = v(\infty)$ and the corresponding Fredholm operators D_u and D_v are surjective. This means that the moduli spaces $\mathcal{M}(A, J)$ and $\mathcal{M}(B, J)$ are smooth manifolds near u and v, respectively. However, as pointed out to us by Gang Liu, this condition alone cannot be enough to obtain a nearby J-holomorphic curve w with surjective Fredholm operator D_w. Consider for example the case where $c_1(A) < 0$ and $c_1(B) < 0$ with $\operatorname{index} D_u = 2n + 2c_1(A) > 0$ and $\operatorname{index} D_v = 2n + 2c_1(B) > 0$ but $\operatorname{index} D_w = 2n + 2c_1(A + B) < 0$. Then there cannot be any J-holomorphic curve in the class $A + B$ with surjective Fredholm operator. To obtain such a curve we must assume that

$$2n + 2c_1(A) + 2c_1(B) > 0.$$

In fact we shall assume that the evaluation map

$$\mathcal{M}(A, J) \times \mathcal{M}(B, J) \to M \times M : (u, v) \mapsto (u(0), v(\infty))$$

is transverse to the diagonal in $M \times M$. This implies the above inequality. Moreover, by Theorem 6.3.2, this is satisfied for generic J.

Given two curves $u \in \mathcal{M}(A, J)$ and $v \in \mathcal{M}(B, J)$ with $u(0) = v(\infty)$ we shall then consider an approximate J-holomorphic curve $u \# v$ and prove that nearby there is an actual J-holomorphic curve w. In particular, this involves a proof that the Fredholm operator $D_{u \# v}$ is onto and an estimate for the right inverse. For this we shall need the following elementary, but subtle, observations about cutoff functions. Throughout the appendix all integrals are to be understood with respect to the Lebesgue measure unless otherwise mentioned.

A.1 Cutoff functions

Lemma A.1.1 *For every constant $\varepsilon > 0$ there exists a $\delta > 0$ and a smooth cutoff function $\beta : \mathbb{R}^2 \to [0,1]$ such that*

$$\beta(z) = \begin{cases} 1, & \text{if } |z| \leq \delta, \\ 0, & \text{if } |z| \geq 1, \end{cases}$$

and

$$\int_{|z| \leq 1} |\nabla \beta(z)|^2 \leq \varepsilon.$$

In fact, such a function exists with $\delta = e^{-2\pi/\varepsilon}$.

Proof: We first define a cutoff function of class $W^{1,2}$ by $\beta(z) = 1$ for $|z| \leq \delta$, $\beta(z) = 0$ for $|z| \geq 1$, and

$$\beta(z) = \frac{\log |z|}{\log \delta}, \qquad \delta \leq |z| \leq 1.$$

Then the gradient $\nabla \beta$ satisfies

$$|\nabla \beta(z)| = \frac{1}{|z| \cdot |\log \delta|}.$$

Hence

$$\int_{\delta \leq |z| \leq 1} |\nabla \beta(z)|^2 = \int_{\delta \leq |z| \leq 1} \frac{1}{|z|^2 |\log \delta|^2} = \int_{\delta}^{1} \frac{2\pi}{s |\log \delta|^2} \, ds = \frac{2\pi}{|\log \delta|}.$$

This function β is only of class $W^{1,2}$ but not smooth. To obtain a smooth function take the convolution with $\phi_N(z) = N^2 \phi(Nz)$ where N is large and $\phi : \mathbb{R}^2 \to \mathbb{R}$ is any smooth function with support in the unit ball and mean value 1. $\qquad \square$

The previous lemma says that there is a sequence of compactly supported functions on $\mathbb{R}^2$ which converge to zero in the $W^{1,2}$ norm but not in the L^∞-norm. This is a borderline case for the Sobolev estimates. For $p > 2$ there is an embedding $W^{1,p}(\mathbb{R}^2) \hookrightarrow C(\mathbb{R}^2)$ and hence the assertion of the previous lemma does not hold if we replace the L^2-norm of $\nabla \beta$ by the L^p-norm for $p > 2$. In the following we shall see that the L^2-norm is precisely what is needed for the proof of surjectivity of a first order operator in 2 dimensions. A similar argument in n dimensions requires the L^n-norm of $\nabla \beta$ to be small and this is again a Sobolev borderline case. The case $n = 4$ is relevant for the gluing of anti-self-dual instantons and this is explained in [13], page 287–295.

Now consider the functions

$$\beta_\lambda(z) = \beta(\lambda z)$$

where β is as in Lemma A.1.1 and $\lambda \geq 1$. In view of the conformal invariance of the L^2-norm of $\nabla \beta$ we have the following estimate for $q < 2$ and $\xi \in W^{1,2}(\mathbb{R}^2)$

$$\|\nabla \beta_\lambda \cdot \xi\|_{L^q} \leq \|\nabla \beta_\lambda\|_{L^2} \|\xi\|_{L^r} \leq c \|\nabla \beta_\lambda\|_{L^2} \|\xi\|_{W^{1,q}}.$$

The first inequality above uses the Hölder inequality

$$\left(\int |uv|^q\right)^{1/q} \leq \left(\int |u|^s\right)^{1/s} \left(\int |v|^r\right)^{1/r}, \qquad \frac{1}{s} + \frac{1}{r} = \frac{1}{q},$$

with $s = 2$. Hence we must take $r = 2q/(2-q)$ and so, by Theorem B.1.5, there is a Sobolev embedding $W^{1,q} \hookrightarrow L^r$. This implies the second inequality above. If the L^2-norm of $\nabla\beta$ is sufficiently small then we obtain

$$\|\nabla\beta_\lambda \cdot \xi\|_{L^q} \leq \varepsilon \|\xi\|_{W^{1,q}}$$

for every $\xi \in W^{1,q}(\mathbb{R}^2)$. In the above proof of this inequality it is essential to assume that $q < 2$. However, the following lemma shows that the last inequality remains valid for $q > 2$ provided that $\xi(0) = 0$.

Lemma A.1.2 *For every $p > 2$ and every $\varepsilon > 0$ there exists a $\delta > 0$ and a smooth cutoff function $\beta : \mathbb{R}^2 \to [0,1]$ as in Lemma A.1.1 such that*

$$\|\nabla\beta_\lambda \cdot \xi\|_{L^p} \leq \varepsilon \|\xi\|_{W^{1,p}}$$

for every $\xi \in W^{1,p}(\mathbb{R}^2)$ with $\xi(0) = 0$ and every $\lambda \geq 1$. Here $\beta_\lambda : \mathbb{R}^2 \to [0,1]$ is defined by $\beta_\lambda(z) = \beta(\lambda z)$.

Proof: With β as in the proof of Lemma A.1.1

$$|\nabla\beta_\lambda(z)| = \lambda|\nabla\beta(\lambda z)| = \frac{1}{|z| \cdot |\log\delta|}.$$

Moreover, by Theorem B.1.4, ξ is Hölder continuous with exponent $\mu = 1 - 2/p$ and hence there exists a constant $c > 0$ (depending only on p) such that

$$|\xi(z)| \leq c|z|^{1-2/p} \|\xi\|_{W^{1,p}}$$

for every $\xi \in W^{1,p}(\mathbb{R}^2)$ with $\xi(0) = 0$. Hence

$$\begin{aligned}
\int_{\delta/\lambda \leq |z| \leq 1/\lambda} |\nabla\beta_\lambda(z)|^p |\xi(z)|^p &\leq \int_{\delta/\lambda \leq |z| \leq 1/\lambda} \frac{1}{|z|^p |\log\delta|^p} c^p |z|^{p-2} \|\xi\|_{W^{1,p}}^p \\
&\leq c^p \int_{\delta/\lambda \leq |z| \leq 1/\lambda} \frac{1}{|z|^2 |\log\delta|^p} \|\xi\|_{W^{1,p}}^p \\
&\leq c^p \int_{\delta/\lambda}^{1/\lambda} \frac{2\pi}{s |\log\delta|^p} \, ds \, \|\xi\|_{W^{1,p}}^p \\
&= c^p \frac{2\pi}{|\log\delta|^{p-1}} \|\xi\|_{W^{1,p}}^p .
\end{aligned}$$

This proves the lemma with $\beta \in W^{1,\infty}$. To obtain a smooth cutoff function one uses the usual convolution argument. $\qquad\square$

A.2 Connected sums of J-holomorphic curves

Fix a regular almost complex structure J such that the operator D_u is onto for every J-holomorphic curve $u : \mathbb{C}P^1 \to M$. Consider the space

$$\mathcal{M}(A, B, J) \subset \mathcal{M}(A, J) \times \mathcal{M}(B, J)$$

of intersecting pairs (u, v) of J-holomorphic curves which represent the classes $A, B \in H_2(M)$ and satisfy

$$u(0) = v(\infty).$$

Denote by $\mathcal{M}_K(A, B, J) \subset \mathcal{M}(A, B, J)$ the subset of those pairs (u, v) which satisfy $\|du\|_{L^\infty} \le K$ and $\|dv\|_{L^\infty} \le K$, where the norm is taken with respect to the Fubini-Study metric on $\mathbb{C}P^1$ and the J-induced metric (3.1) on M.

Given a pair $(u, v) \in \mathcal{M}_K(A, B, J)$ and a large number $R > 0$ we shall construct an approximate J-holomorphic curve $w_R : \mathbb{C}P^1 \to M$ which (approximately) agrees with u on the complement of a disc $B_{1/R}$ of radius $1/R$ and with the rescaled curve $v(R^2 z)$ on $B_{1/R}$. The condition $u(0) = v(\infty)$ guarantees that these maps are approximately equal on the circle $|z| = 1/R$. We shall thicken this circle on each side to annuli of the form $A(\delta r, r)$ for a sufficiently small number $\delta > 0$. We shall use cutoff functions as constructed in Lemma A.1.1 which are supported in these annuli and this determines the smallness of δ. We shall fix a suitable constant δ and consider sufficiently large numbers $R > R_\delta$ so that the product δR is large. In the limit $R \to \infty$ we obtain the converse of Gromov's compactness. In other words the curves w_R will converge in the weak sense of Section 4.4 to the pair (u, v).

More precisely, for any sufficiently small number $\delta > 0$, any sufficiently large number $R > 0$, and any pair $(u, v) \in \mathcal{M}_K(A, B, J)$ we shall construct an *approximate J-holomorphic curve* $w_R = u \#_R v : \mathbb{C}P^1 \to M$ which satisfies

$$
w_R(z) = \begin{cases}
v(R^2 z), & \text{if } |z| \le \frac{\delta}{2R}, \\[2mm]
u(0) = v(\infty), & \text{if } \frac{\delta}{R} \le |z| \le \frac{1}{\delta R}, \\[2mm]
u(z), & \text{if } |z| \ge \frac{2}{\delta R}.
\end{cases}
$$

To define the function w_R in the rest of the annulus $\delta/2R \le |z| \le 2/\delta R$ we fix a cutoff function $\rho : \mathbb{C} \to [0, 1]$ such that

$$
\rho(z) = \begin{cases}
1, & \text{if } |z| \ge 2, \\
0, & \text{if } |z| \le 1.
\end{cases}
$$

Now denote the point of intersection of the curves u and v by $x = u(0) = v(\infty)$ and use the exponential map in a neighbourhood of this point. Let $\xi_u(z) \in T_x M$ for $|z| < \varepsilon$ and $\xi_v(z) \in T_x M$ for $|z| > 1/\varepsilon$ be the vector fields such that $u(z) = \exp_x(\xi_u(z))$ and $v(z) = \exp_x(\xi_v(z))$. Define

$$w_R(z) = \exp_x\left(\rho(\delta R z)\xi_u(z) + \rho(\delta R/z)\xi_v(R^2 z)\right)$$

for $\delta/2R \le |z| \le 2/\delta R$. This is well defined whenever $R > 2/\delta\varepsilon$. Moreover, the number $\varepsilon > 0$ depends only on the uniform bound K for du and dv. The map w_R is not a J-holomorphic curve, however we shall see that the error converges to zero as $R \to \infty$. Our goal is construct a *true J-holomorphic curve* $\tilde{w}_R$ near w_R.

Remark A.2.1 The curves w_R converge to the pair (u, v) in the weak sense of Section 4.4. More precisely, $w_R(z)$ converges to $u(z)$ uniformly with all derivatives on compact subsets of $\mathbb{C}P^1 - \{0\}$ and $w_R(z/R^2)$ converges to $v(z)$ uniformly with all derivatives on compact subsets of $\mathbb{C} = \mathbb{C}P^1 - \{\infty\}$. The J-holomorphic curves $\widetilde{w}_R$ which we construct below will converge to the pair (u, v) in the same way. □

Remark A.2.2 Consider the curves $u_R, v_R : \mathbb{C}P^1 \to M$ defined by

$$u_R(z) = \begin{cases} w_R(z), & \text{if } |z| \geq 1/R, \\ u(0), & \text{if } |z| \leq 1/R, \end{cases}$$

and

$$v_R(z) = \begin{cases} w_R(z/R^2), & \text{if } |z| \leq R, \\ v(\infty), & \text{if } |z| \geq R. \end{cases}$$

Then u_R converges to u in the $W^{1,p}$-norm and v_R converges to v in the $W^{1,p}$-norm. Moreover, the convergence is uniform over all pairs $(u, v) \in \mathcal{M}_K(A, B, J)$. Note that uniformity requires the uniform estimate on the derivatives of du and dv (by the constant K). Note also that u_R and v_R do not converge in the C^1-norm. □

Example A.2.3 Consider the case $M = \mathbb{C}P^1$ with the standard complex structure. Then the holomorphic curves

$$u(z) = 1 + z, \qquad v(z) = 1 + 1/z$$

satisfy $u(0) = v(\infty) = 1$ as required. The above maps $w_R : \mathbb{C}P^1 \to \mathbb{C}P^1$ satisfy

$$w_R(z) = \begin{cases} 1 + 1/R^2 z, & \text{if } |z| \leq \delta/2R, \\ 1, & \text{if } \delta/R \leq |z| \leq 1/\delta R, \\ 1 + z, & \text{if } |z| \geq 2/\delta R. \end{cases}$$

Nearby J-holomorphic curves are given by

$$\widetilde{w}_R(z) = v(R^2 z) + u(z) - 1 = \frac{z^2 + z + 1/R^2}{z}$$

and these converge to the pair (u, v) as in Remark A.2.1. □

A.3 Weighted norms

In order to apply the implicit function theorem we must specify the norm in which $\bar{\partial}_J(w_R)$ is small. The guiding principle for our choice of norm is the observation that the curves u and v play equal roles in this gluing argument. However, the map v appears in rescaled form and is concentrated in a small ball of radius $1/R$. So in order to give u and v equal weight we shall consider a family of R-dependent metrics on the 2-sphere such that the volume of the ball of radius $1/R$ is approximately equal to the volume of S^2 with respect to the standard (Fubini-Study) metric. The rescaled metric is of the form $\theta_R^{-2}(ds^2 + dt^2)$ where

$$\theta_R(z) = \begin{cases} R^{-2} + R^2|z|^2, & \text{if } |z| \leq 1/R, \\ 1 + |z|^2, & \text{if } |z| \geq 1/R. \end{cases}$$

The area of S^2 with respect to this metric is given by

$$\mathrm{Vol}_R(S^2) = \int_{\mathbb{C}} \theta_R^{-2} \leq 2\pi. \tag{A.1}$$

where, as always, we integrate with respect to Lebesgue measure. The effect of this metric on the L^p and $W^{1,p}$ norms of vector fields along w_R is as follows. If we rescale the vector field $\xi(z) \in T_{w_R(z)}M$ with support in $B_{1/2R}$ to obtain a vector field $\xi(z/R^2) \in T_{v(z)}M$ along v then the standard norms of this rescaled vector field agree with the weighted norms of the original vector field ξ. More explicitly, define the weighted norms

$$\|\xi\|_{0,p,R} = \left(\int_{\mathbb{C}} \theta_R(z)^{-2} |\xi(z)|^p \right)^{1/p},$$

$$\|\xi\|_{1,p,R} = \left(\int_{\mathbb{C}} \theta_R(z)^{-2} |\xi(z)|^p + \theta_R(z)^{p-2} |\nabla \xi(z)|^p \right)^{1/p}.$$

Here ∇ denotes the Levi-Civita connection of the metric induced by J and, with $z = s + it$, we denote $|\nabla \xi(z)|^2 = |\nabla_s \xi(z)|^2 + |\nabla_t \xi(s)|^2$. Similarly, for 1-forms $\eta = \eta_1 ds + \eta_2 dt \in \Omega^1(w_R^* TM)$, consider the norms

$$\|\eta\|_{0,p,R} = \left(\int_{\mathbb{C}} \theta_R(z)^{p-2} |\eta(z)|^p \right)^{1/p},$$

$$\|\eta\|_{0,\infty,R} = \sup_{z \in \mathbb{C}} \theta_R(z) |\eta(z)|,$$

$$\|\eta\|_{1,p,R} = \left(\int_{\mathbb{C}} \theta_R(z)^{p-2} |\eta(z)|^p + \theta_R(z)^{2p-2} |\nabla \eta(z)|^p \right)^{1/p}.$$

where $|\eta|^2 = |\eta_1|^2 + |\eta_2|^2$ and $|\nabla \eta|^2 = |\nabla_s \eta_1|^2 + |\nabla_t \eta_1|^2 + |\nabla_s \eta_2|^2 + |\nabla_t \eta_2|^2$.

In the case $R = 1$ these are the usual L^p and $W^{1,p}$ norms with respect to the Fubini-Study metric on $\mathbb{C}$ (as a coordinate patch of $\mathbb{C}P^1$). For general R these norms should be considered in two parts. In the domain $|z| \geq 1/R$ they are still the usual norms and in the domain $|z| \leq 1/R$ they agree with the usual norms of the rescaled vector field

$$\widetilde{\xi}(z) = \xi(R^{-2} z)$$

or 1-form

$$\widetilde{\eta} = R^{-2}(\eta_1(R^{-2}z)ds + \eta_2(z/R^2)dt)$$

along $v(z)$ in the ball of radius R, again with respect to the Fubini-Study metric on $\mathbb{C}$. Hence we obtain the usual Sobolev estimates (see Theorems B.1.4 and B.1.5) with constants which are independent of R.

Lemma A.3.1 (i) *For $p > 2$ and $K > 0$ there exist constants $R_0 > 0$ and $c > 0 such that*

$$\|\xi\|_{L^\infty} \leq c \|\xi\|_{1,p,R}, \qquad \|\eta\|_{0,\infty,R} \leq c \|\eta\|_{1,p,R}$$

for all $R \geq R_0$, $\xi \in C^\infty(w_R^* TM)$, $\eta \in \Omega^1(w_R^* TM)$, $(u,v) \in \mathcal{M}_K(A, B, J)$.

(ii) *For $p < 2$, $1 \leq q \leq 2p/(2-p)$, and $K > 0$ there exist constants $R_0 > 0$ and $c > 0$ such that*

$$\|\xi\|_{0,q,R} \leq c \|\xi\|_{1,p,R}, \qquad \|\eta\|_{0,q,R} \leq c \|\eta\|_{1,p,R}$$

for all $R \geq R_0$, $\xi \in C^\infty(w_R^* TM)$, $\eta \in \Omega^1(w_R^* TM)$, $(u,v) \in \mathcal{M}_K(A,B,J)$.

The following lemma can be proved by a straightforward argument involving change of variables. It plays a key role in our application of the implicit function theorem. Note in fact that in Theorem 3.3.4 we do not specify the metric on the Riemann surface $\Sigma = \mathbb{C}P^1$ with respect to which the norms are defined. We do however assume that the volume and the L^p-norm of du are uniformly bounded with respect to this metric.

Lemma A.3.2 *For $p > 2$ and $K > 0$ there exist constants $R_0 > 0$ and $c > 0$ such that*

$$\|dw_R\|_{0,p,R} \leq c$$

for $R \geq R_0$ and $(u,v) \in \mathcal{M}_K(A,B,J)$.

A.4 An estimate for the inverse

We must now prove that the Fredholm operator D_{w_R} is surjective with a uniformly bounded right inverse provided that $R > 0$ is sufficiently large. This argument will involve the cutoff function of Lemma A.1.1. We use the notation of Section 3.3. We begin by rephrasing our transversality condition. Abbreviate

$$W_{u,v}^{1,p} = \left\{ (\xi_u, \xi_v) \in W^{1,p}(u^*TM) \times W^{1,p}(v^*TM) \mid \xi_u(0) = \xi_v(\infty) \right\}$$

and

$$L_u^p = L^p(\Lambda^{0,1}T^*\mathbb{C}P^1 \otimes_J u^*TM).$$

Note that this definition of the space $W_{u,v}^{1,p}$ only makes sense for $p > 2$ since it is only in this case that $W^{1,p}$-sections are continuous and can be evaluated at a point. The space $W_{u,v}^{1,p}$ should be interpreted as a limit as $R \to \infty$ of the spaces $W^{1,p}(w_R^* TM)$ of vector fields along the glued curves. Our transversality assumption means that the operator

$$D_{u,v} : W_{u,v}^{1,p} \to L_u^p \times L_v^p, \qquad D_{u,v}(\xi_u, \xi_v) \mapsto (D_u\xi_u, D_v\xi_v)$$

is onto. Hence $D_{u,v}$ has a right inverse $Q_{u,v} : L_u^p \times L_v^p \to W_{u,v}^{1,p}$ such that

$$D_{u,v} \circ Q_{u,v} = \mathbb{1}, \qquad \|Q_{u,v}\| \leq c_0.$$

Here the constant c_0 can be chosen independent of the pair $(u,v) \in \mathcal{M}_K(A,B,J)$ but will in general depend on A, B, K, and J. Because the space $\mathcal{M}_K(A,B,J)$ is compact it suffices to prove this locally and this can be done by reducing the dimension of the kernel of $D_{u,v}$ to zero through additional (say pointwise) conditions on the pair (ξ_u, ξ_v).

Remark A.4.1 Here are a few more comments on the existence of the uniformly bounded inverse $Q_{u,v}$ of the operator $D_{u,v}$.

1. Assume first that $D_{u,v}$ has index zero. Then it is bijective and the open mapping theorem shows that $Q_{u,v} = D_{u,v}^{-1}$ is bounded. Moreover, the operator $D_{u,v}$ depends continuously on (u,v) in the norm topology. This implies that also the operator $Q_{u,v}$ depends continuously on (u,v) in the norm topology and hence the required estimate is uniform in the case of index zero. (Warning: Note that the domains of the operators and their range actually depend on the pair (u,v). So one has to choose some kind of identification of say, $W_{u,v}^{1,p}$ and $W_{u',v'}^{1,p}$ for nearby pairs (u,v) and (u',v') to make this continuous dependence precise. This can be done by parallel transport along short geodesics as in the proof of Theorem 3.3.4.)

2. If the index is bigger than zero, for example $2n$, one can fix some point $z_0 \in S^2$ and impose the condition $\xi_u(z_0) = 0$. For a generic point z_0 this cuts down the dimension of the kernel by $2n$ and results in an operator of index zero, reducing to the previous case. More generally cut down by imposing the condition $\xi_u(z_0) \in E_{u(z_0)}$ for some subbundle $E \subset TM$ of the right dimension. In any case, reduce to the case of index zero to deal with the general case. $\quad\square$

Our strategy is now to use the operator $Q_{u,v}$ to construct, for R sufficiently large, an approximate right inverse $Q_R : L_{w_R}^p \to W^{1,p}(w_R^*TM)$ of the operator D_{w_R} such that

$$\|Q_R\| \le c_1, \qquad \|D_{w_R}Q_R - \mathbb{1}\| < 1/2. \tag{A.2}$$

Under these conditions the operator $D_{w_R}Q_R : L_{w_R}^p \to L_{w_R}^p$ is invertible and a right inverse of D_{w_R} is given by

$$Q_{w_R} = Q_R(D_{w_R}Q_R)^{-1}.$$

Roughly speaking, we will construct the operator Q_R by means of the commutative diagram

$$\begin{array}{ccc} W_{u,v}^{1,p} & \xleftarrow{Q_{u,v}} & L_u^p \times L_v^p \\ \downarrow & & \uparrow \\ W^{1,p}(w_R^*TM) & \xleftarrow{Q_R} & L_{w_R}^p. \end{array} \tag{A.3}$$

Here the vertical maps are given in terms of cutoff functions. In fact it is convenient to modify the diagram slightly and replace u and v by the curves u_R and v_R defined in Remark A.2.2. Then u_R converges to u in the $W^{1,p}$-norm and similarly for v_R. Hence the operator D_{u_R,v_R} still has a uniformly bounded right inverse Q_{u_R,v_R}. The following diagram should serve as a guide to the definition of Q_R.

$$\begin{array}{ccc} (\xi_u,\xi_v) & \xleftarrow{Q_{u_R,v_R}} & (\eta_u,\eta_v) \\ \downarrow & & \uparrow \\ \xi & \xleftarrow{Q_R} & \eta. \end{array}$$

Given $\eta \in L_{w_R}^p$ we define the pair $(\eta_u,\eta_v) \in L_u^p \times L_v^p)$ simply by cutting off η along the circle $|z| = 1/R$:

$$\eta_u(z) = \begin{cases} \eta_R(z), & \text{if } |z| \ge 1/R, \\ 0, & \text{if } |z| \le 1/R, \end{cases} \qquad \eta_v(z) = \begin{cases} R^{-2}\eta_R(R^{-2}z), & \text{if } |z| \le R, \\ 0, & \text{if } |z| \ge R. \end{cases}$$

The discontinuities in η_u and η_v do not cause problems because only their L^p-norms enter the estimates. Now the pair (ξ_u, ξ_v) is defined in terms of the right inverse Q_{u_R, v_R} by

$$(\xi_u, \xi_v) = Q_{u_R, v_R}(\eta_u, \eta_v).$$

In particular

$$\xi_u(0) = \xi_v(\infty) = \xi_0 \in T_{u(0)}M.$$

Finally, we define $\xi = Q_R \eta$ by

$$\xi(z) = \begin{cases} \xi_u(z), & \text{if } |z| \geq \frac{1}{\delta R}, \\[2mm] \xi_u(z) + (1 - \beta(1/Rz))(\xi_v(R^2 z) - \xi_0), & \text{if } \frac{1}{R} \leq |z| \leq \frac{1}{\delta R}, \\[2mm] \xi_v(R^2 z) + (1 - \beta(Rz))(\xi_u(z) - \xi_0), & \text{if } \frac{\delta}{R} \leq |z| \leq \frac{1}{R}, \\[2mm] \xi_v(R^2 z), & \text{if } |z| \leq \frac{\delta}{R}. \end{cases}$$

The easiest way to understand this definition is by considering the case $\xi_0 = 0$. Then in the annulus $\delta/R \leq |z| \leq 1/\delta R$ the maps u_R, v_R and w_R all take the constant value x, and the function $\xi(z) \in T_x M$ is simply the superposition of the functions $(1 - \beta(Rz))\xi_u(z)$ and $(1 - \beta(1/Rz))\xi_v(R^2 z)$. The important fact is that, in the first term, the cutoff function $\beta(Rz)$ only takes effect in the region $|z| \leq 1/R$ where $D_{u_R}\xi_u = 0$. Similarly for the second term because the construction is completely symmetric in u and v. In the case $\xi_0 \neq 0$ the formula can be interpreted in the same way, but relative to the "origin" ξ_0.

Exercise A.4.2 Check the symmetry in the formula for ξ by considering the functions $\widetilde{u}(z) = v(1/z)$, $\widetilde{v}(z) = u(1/z)$, $\widetilde{w}_R(z) = w_R(1/R^2 z)$ and the vector fields $\widetilde{\xi}_u(z) = \xi_v(1/z)$, $\widetilde{\xi}_v(z) = \xi_u(1/z)$, $\widetilde{\xi}(z) = \xi(1/R^2 z)$.

Lemma A.4.3 *The operator Q_R satisfies the estimate (A.2).*

Proof: We must prove that

$$\|D_{w_R}\xi - \eta\|_{0,p,R} \leq \frac{1}{2} \|\eta\|_{0,p,R}. \tag{A.4}$$

Since $D_{u_R}\xi_u = \eta_u$ and $D_{v_R}\xi_v = \eta_v$ the term on the left hand side vanishes for $|z| \geq 1/\delta R$ and for $|z| \leq \delta/R$. In view of the symmetry of our formulae it suffices to estimate the left hand side in the annulus

$$\frac{\delta}{R} \leq |z| \leq \frac{1}{R}.$$

In this region, $u_R = v_R = w_R$ is the constant map. Therefore over this annulus the corresponding operators D_{u_R}, D_{v_R}, and D_{w_R} are all equal, and on functions are simply the usual $\bar{\partial}$-operator. Further, the definition of ξ_v implies that $D_{w_R}\xi_v(R^2 \cdot) = \eta$. Hence, with the notation $\beta_R(z) = \beta(Rz)$,

$$\begin{aligned} D_{w_R}\xi - \eta &= D_{u_R}((1 - \beta_R)(\xi_u - \xi_0)) \\ &= (1 - \beta_R)D_{u_R}(\xi_u - \xi_0) + \bar{\partial}\beta_R \otimes (\xi_u - \xi_0) \\ &= (\beta_R - 1)D_{u_R}\xi_0 + \bar{\partial}\beta_R \otimes (\xi_u - \xi_0). \end{aligned}$$

Here we have used the crucial fact that $D_{u_R}\xi_u = \eta_u = 0$ in the region $|z| \leq 1/R$.

Now we must estimate the L^p-norm of this 1-form with respect to the R-dependent metric. The next crucial point to observe is that the weighting function for 1-forms is $\theta_R(z)^{p-2}$. Since $p > 2$ and $\theta_R(z) \leq \theta_1(z) \leq 2$ in the region $|z| \leq 1/R$, it follows that the $(0, p, R)$-norm of our 1-form is smaller that the ordinary L^p-norm (up to a universal factor less than 2). Hence we obtain the inequality

$$
\begin{aligned}
\|D_{w_R}\xi - \eta\|_{0,p,R;B_{1/R}} \;&\leq\; 2\,\|D_{w_R}\xi - \eta\|_{L^p(B_{1/R})} \\
&\leq\; 2\,\|D_{u_R}\xi_0\|_{L^p(B_{1/R})} + 2\,\|\bar\partial\beta_R \otimes (\xi_u - \xi_0)\|_{L^p(B_{1/R})} \\
&\leq\; c_2 R^{-2/p}|\xi_0| + \varepsilon\,\|\xi_u - \xi_0\|_{W^{1,p}} \\
&\leq\; c_3(\varepsilon + R^{-2/p})\left(\|\eta_u\|_{L^p} + \|\eta_v\|_{L^p}\right) \\
&=\; c_3(\varepsilon + R^{-2/p})\,\|\eta\|_{0,p,R}
\end{aligned}
$$

The third inequality follows from the fact that the term $D_{u_R}\xi_0$ can be pointwise estimated by $|\xi_0|$ and the L^p-norm is taken over an area at most π/R^2. Moreover, we have used Lemma A.1.2. The fourth inequality follows from the uniform estimate for the right inverse Q_{u_R,v_R} of D_{u_R,v_R}. The last equality follows from the definition of η_u and η_v.

In view of the symmetry, spelled out in Exercise A.4.2, we have a similar estimate in the domain $|z| \geq 1/R$ and this proves that the operator Q_R satisfies the second inequality in (A.2) provided that $R \geq R_\delta$ is sufficiently large. The first inequality is an easy exercise which can be safely left to the reader. $\qquad\square$

Lemma A.4.4 *For every constant $1 \leq p < \infty$ there exist constants $R_0 > 0$ and $c > 0$ such that*

$$
\left\|\bar\partial_J(w_R)\right\|_{0,p,R} \leq c(\delta R)^{-2/p}
$$

for $R \geq R_\delta$. Moreover, given a number $K > 0$ we may choose the same constants R_δ and c for all pairs $(u, v) \in \mathcal{M}_K(A, B, J)$.

Proof: Use the fact that $w_R(z) = u_R(z)$ for $|z| \geq 1/R$ and that, in view of Remark A.2.2, u_R converges to the J-holomorphic curve u in the $W^{1,p}$-norm. For $|z| \leq 1/R$ one can again use the symmetry of Exercise A.4.2. To obtain the more precise estimate of the convergence rate recall that $u_R(z) = \exp_x(\rho(\delta Rz)\xi_u(z))$. Hence the term $\bar\partial_J(w_R)$ is supported in the annulus $B_{2/\delta R} - B_{1/\delta R}$ and in this region can be expressed as a sum of two terms. One involves the first derivatives of the cutoff function $\rho(\delta Rz)$, which grow like δR, multiplied by the function $\xi_u(z)$ which is bounded by a constant times $1/\delta R$ provided that $|z| \leq 2/\delta R$. The second term involves the first derivatives of $\xi_u(z)$ and these are again uniformly bounded. Since the integral is over a region of area at most $4\pi(\delta R)^{-2}$ this proves the lemma. $\square$

A.5 Gluing

The Lemmata A.3.2, A.4.4 and A.4.3 imply that for R sufficiently large the smooth map $w_R : \mathbb{C}P^1 \to M$ satisfies the requirements of the implicit function theorem 3.3.4. Hence it follows from Theorem 3.3.4 that, for $R > 0$ sufficiently large, there exists a unique smooth section $\xi_R = Q_{w_R}\eta_R \in C^\infty(w_R^* TM)$ such that the map

$$\widetilde{w}_R(z) = \exp_{w_R(z)}(\xi_R(z))$$

is a J-holomorphic curve and

$$\|\xi_R\|_{W^{1,p}} \le cR^{-2/p}.$$

Thus we have constructed, for every sufficiently large real number R, a smooth map

$$f_R : \mathcal{M}_K(A, B, J) \to \mathcal{M}(A + B, J)$$

given by

$$f_R(u, v) = \widetilde{w}_R.$$

This construction works for $R > R_0$ where R_0 depends on A, B, and K. The following lemma makes precise the sense in which the above gluing construction is the converse of Gromov compactness. The proof is obvious, except perhaps for the last statement which can be verified with the methods in the proof of Theorem 4.4.3.

Lemma A.5.1 *As* $R \to \infty$ *the* J*-holomorphic curves* $\widetilde{w}_R = f_R(u, v)$ *converge weakly to the pair* (u, v) *in the sense of Section 4.4. More precisely,* $\widetilde{w}_R$ *converges on compact subsets of* $\mathbb{C}P^1 - \{0\}$ *to* u, $\widetilde{w}_R(R^2 z)$ *converges on compact subsets of* $\mathbb{C} = \mathbb{C}P^1 - \{\infty\}$ *to* v, *and*

$$|z_R| \to 0, \quad |z_R|R^2 \to \infty \quad \implies \quad \widetilde{w}_R(z_R) \to u(0) = v(\infty).$$

We shall now discuss the properties of the gluing map f_R for a fixed value of R. Recall from the discussion after Theorem 5.2.1 that the space $\mathcal{M}(A, B, J)$ is a smooth manifold whenever the evaluation map

$$\mathcal{M}(A, J) \times \mathcal{M}(B, J) \to M^2 : (u, v) \mapsto (u(0), v(\infty))$$

is transverse to the diagonal. Moreover its dimension is dim $\mathcal{M}(A, B, J) = 2n + 2c_1(A + B)$ and agrees with that of $\mathcal{M}(A + B, J)$. It follows from the uniqueness result in Proposition 3.3.5 that f_R is a diffeomorphism between open sets in $\mathcal{M}(A, B, J)$ and $\mathcal{M}(A + B, J)$. Moreover, the map f_R is orientation preserving. To see this examine the proof of Lemma A.4.3 and use a similar cutoff argument to find an isomorphism

$$F_R(u, v) : \{(\xi_u, \xi_v) \in \ker D_u \oplus \ker D_v \mid \xi_u(0) = \xi_v(\infty)\} \to \ker D_{w_R}.$$

This linear isomorphism mimics the nonlinear gluing construction. One first uses a cutoff function to obtain a section of the bundle $w_R^* TM$ which is *approximately* in the kernel of D_{w_R} and then projects orthogonally onto the kernel of D_{w_R}. The resulting isomorphism represents (approximately) the differential of the map f_R at the point (u, v) and it can easily be seen to be orientation preserving. We summarize our findings in the following

Theorem A.5.2 *The above map* $f_R : \mathcal{M}_K(A, B, J) \to \mathcal{M}(A + B, J)$ *is an orientation preserving diffeomorphism onto an open set in* $\mathcal{M}(A + B, J)$.

Remark A.5.3 Note that the image of f_R has compact closure in $\mathcal{M}(A + B, J)$ and so is not a full neighbourhood of the boundary cusp-curves. The situation may be explained as follows. There is an 8-dimensional reparametrization group acting on $\mathcal{M}(A, B, J)$ and a 6-dimensional group acting on $\mathcal{M}(A+B, J)$. In $\mathcal{M}_K(A, B, J)$ we can think of the variation of the points 1 and ∞ in the domain of u and the variation of 0 in the domain of v as corresponding to this 6-dimensional group acting on $\mathcal{M}(A + B, J)$. Theorem A.5.2 shows that the "extra" variation of 1 in the domain of v is translated by f_R into a 2-dimensional family of geometrically distinct $(A+B)$-curves. Thus for each pair $(u, v) \in \mathcal{M}_K(A, B, J)$ the map f_R takes the annulus

$$\{(u, v \circ \phi) \in \mathcal{M}_K(A, B, J) \mid \phi(0) = 0,\ \phi(\infty) = \infty\}$$

diffeomorphically onto an annulus in the space of unparametrized $(A + B)$-curves, which surrounds the cusp-curve (u, v) in the same way that a large annulus $\{r_1 < |z| < r_2\}$ surrounds ∞ in $\mathbb{C}$. Here ϕ plays the role of Floer's gluing parameter.

Proof of Lemma 8.2.5: We must prove that for R sufficiently large there is a one-to-one correspondence (preserving intersection numbers) of pairs $(u, v) \in \mathcal{M}(A - B, B, J)$ such that

$$u(\infty) \in \alpha_1, \quad u(1) \in \alpha_2, \quad v(0) \in \alpha_3, \quad v(1) \in \alpha_4, \tag{A.5}$$

with curves $w \in \mathcal{M}(A, J)$ such that

$$w(\infty) \in \alpha_1, \quad w(1) \in \alpha_2, \quad w(0) \in \alpha_3, \quad w(1/R^2) \in \alpha_4. \tag{A.6}$$

The strategy is to prove first that every pair (u, v) gives rise to a suitable $w = w_R$ for R sufficiently large and then to show, by a compactness argument, that every w is of this form.

Under our dimension assumption $\sum_j \deg(\alpha_j) = 6n - 2c_1(A)$ the codimension of the pseudo-cycle

$$\alpha = \alpha_1 \times \alpha_2 \times \alpha_3 \times \alpha_4$$

is $2n + 2c_1(A)$ and hence agrees with that of the moduli space $\mathcal{M}(A - B, B, J)$. As we have seen above, there are only finitely many pairs $(u, v) \in \mathcal{M}(A - B, B, J)$ which satisfy (A.5), and so we may choose K so large that all these pairs are contained in $\mathcal{M}_K(A - B, B, J)$. Now consider the 4-fold evaluation map

$$e_R : \mathcal{M}(A, J) \to M^4, \qquad e_R(w) = (w(\infty), w(1), w(0), w(1/R^2))$$

and note that

$$e_R(u \#_R v) = (u(\infty), u(1), v(0), v(1)).$$

By Lemma A.5.1, $f_R(u, v)$ is so close to the curve $u\#_R v$ that the composition

$$e_R \circ f_R : \mathcal{M}_K(A - B, B, J) \to M^4$$

converges in the C^∞-topology to the evaluation map

$$e : \mathcal{M}_K(A - B, B, J) \to M^4, \qquad e(u, v) = (u(\infty), u(1), v(0), v(1)).$$

Since e is transverse to α so is $e_R \circ f_R$ for R sufficiently large, and so $e_R \circ f_R$ must have precisely as many intersection points with α as the limit map e. Thus we have proved that near every solution $(u, v) \in \mathcal{M}(A - B, B, J)$ of (A.5) there is a solution $w \in \mathcal{M}(A, J)$ of (A.6), provided that $R > 0$ is sufficiently large. Since the local diffeomorphism f_R is orientation preserving, the curve w contributes with the same intersection number as (u, v).

Here we have assumed that $0 \neq B \neq A$. If $B = 0$ then v is constant and u represents the class A with

$$u(\infty) \in \alpha_1, \quad u(1) \in \alpha_2, \quad u(0) \in \alpha_3 \cap \alpha_4.$$

Moreover, the evaluation map $e_\infty : \mathcal{M}(A, J) \to M^4$ defined by

$$e_\infty(u) = (u(\infty), u(1), u(0), u(0))$$

is tranverse to $\alpha = \alpha_1 \times \alpha_2 \times \alpha_3 \times \alpha_4$. Since e_R converges to e_∞ in the C^∞-topology we have a nearby curve $w \in \mathcal{M}(A, J)$ with $e_R(w) \in \alpha$. A similar argument applies to the case $B = A$.

Now let (u_j, v_j) for $j = 1, \ldots, N$ denote the finite set of all pairs $(u, v) \in \mathcal{M}(A - B, B, J)$ which satisfy (A.5) for all choices of B, including $B = 0$ and $B = A$. Denote by $w_{j,R} \in \mathcal{M}(A, J)$ the solution of (A.6) which corresponds to the pair (u_j, v_j). We shall use Gromov compactness to show that, for R sufficiently large, every curve $w \in \mathcal{M}(A, J)$ with $e_R(w) \in \alpha$ must be one of the curves $w_{j,R}$. Suppose, by contradiction, that there exists a sequence $R_\nu \to \infty$ and a sequence $w_\nu \in \mathcal{M}(A, J)$ which satisfies (A.6) with $R = R_\nu$ but does not agree with any of the solutions w_{j,R_ν}. Then it follows from Proposition 3.3.5 that there exists a constant $\varepsilon > 0$ such that

$$\sup_{z \in \mathbb{C}P^1} \operatorname{dist}(w_\nu(z), w_{j,R_\nu}(z)) \geq \varepsilon \tag{A.7}$$

for all j and ν. On the other hand, by Gromov's compactness we may assume, passing to a subsequence if necessary, that w_ν converges in the weak sense of Section 4.4 to a cusp curve. Following the reasoning of Section 8.2 we see that the limit curve has only two components and must be one of the pairs (u_j, v_j). Assume first that both components are nonconstant. Then the bubbling must occur at $z = 0$. Hence w_ν converges on compact subsets of $\mathbb{C}P^1 - \{0\}$ to u_j, and converges, after rescaling, on compact subsets of $\mathbb{C} = \mathbb{C}P^1 - \{\infty\}$ to v_j. Examining the values of $w_\nu(0)$ and $w_\nu(1/R_\nu^2)$ we see that the rescaling factor must be R_ν^2. Moreover, the argument in the proof of Theorem 4.4.3 (which shows that $u(0) = v(\infty)$) can be used to prove that

$$|z_\nu| \to 0, \ |z_\nu| R_\nu^2 \to \infty \quad \Longrightarrow \quad w_\nu(z_\nu) \to u_j(0) = v_j(\infty).$$

By construction, the sequence w_{j,R_ν} has exactly the same properties and hence w_ν and w_{j,R_ν} must be arbitrarily close in the C^0-topology provided that ν is sufficiently large. This contradicts (A.7). The extreme cases where either u_j or v_j are constant can be safely left to the reader. This proves the lemma and hence Theorem 8.2.1. $\square$

Remark A.5.4 (i) Lemma 8.2.5 can be proved along a slightly different route, namely by first taking the approximate J-holomorphic curve $w_j = u_j \#_R v_j :$ $\mathbb{C}P^1 \to M$ obtained from the gluing construction. This map satisfies $e_R(w_j) \in$

α and then one can use the infinite dimensional implicit function in the space of all maps w with $e_R(w) \in \alpha$ to find a nearby J-holomorphic curve $\widetilde{w}_j$ with $e_R(\widetilde{w}_j) \in \alpha$. Of course, this produces the same J-holomorphic curves intersecting α, however some additional argument is required to show that the intersection numbers are the same for both invariants $\Psi_A(\alpha_1, \alpha_2, \alpha_3, \alpha_4)$ and $\Psi_{A-B,B}(\alpha_1, \alpha_2; \alpha_3, \alpha_4)$.

(ii) The techniques of this section can obviously be generalized to glue together finitely many intersecting J-holomorphic curves described by a framed class $D = (A_1, \ldots, A_N, j_2, \ldots, j_N)$.

(iii) The map f_R constructed above will in general not preserve any group action. The best way to deal with this problem seems to be to use additional intersection conditions as *gauge fixing* as in (i) above.

(iv) The above techniques can also be used to glue together J-holomorphic maps $u_i : \Sigma_i \to M$ which are defined on Riemann surfaces of higher genus. However, in this case we must allow for the complex structure on the glued Riemann surface $\Sigma = \Sigma_1 \# \Sigma_2$ to vary. The gluing will produce a Riemann surface with a very thin *neck*. Conversely, it follows from Gromov compactness that a sequence $u_\nu : \Sigma \to M$ of (j_ν, J)-holomorphic curves can only split up into two (or more) surfaces of higher genus if the complex structure j_ν on Σ converges to the boundary of Teichmüller space.

Appendix B

Elliptic Regularity

B.1   Sobolev spaces

Throughout $\Omega \subset \mathbb{R}^n$ is an open set with smooth boundary. Denote by $C^\infty(\bar{\Omega})$ the space of restrictions of smooth functions on $\mathbb{R}^n$ to $\bar{\Omega}$ and by $C_0^\infty(\Omega)$ the space of smooth compactly supported functions on Ω.

For a positive integer k and a number $1 \leq p < \infty$ define the $W^{k,p}$-norm of a smooth function $u : \Omega \to \mathbb{R}$ by

$$\|u\|_{k,p} = \left(\int_\Omega \sum_{|\nu| \leq k} |\partial^\nu u(x)|^p \, dx \right)^{1/p}$$

where $\nu = (\nu_1, \ldots, \nu_n)$ is a multi-index and $|\nu| = \nu_1 + \cdots + \nu_n$. The Sobolev space

$$W^{k,p}(\Omega)$$

is defined as the completion of $C^\infty(\bar{\Omega})$ with respect to the $W^{k,p}$-norm. The space

$$W_0^{k,p}(\Omega)$$

is the closure of $C_0^\infty(\Omega)$ in $W^{k,p}(\Omega)$.

The next proposition uses the smoothness of the boundary of Ω to show that every function with *weak L^p-derivatives* up to order k lies in the space $W^{k,p}(\Omega)$.

Proposition B.1.1 *Let $u \in L^p(\Omega)$. Then $u \in W^{k,p}(\Omega)$ if and only if for every $|\nu| \leq k$ there exists a function $u_\nu \in L^p(\Omega)$ such that*

$$\int_\Omega u(x) \partial^\nu \phi(x) \, dx = (-1)^{|\nu|} \int_\Omega u_\nu(x) \phi(x) \, dx$$

for $\phi \in C_0^\infty(\Omega)$.

In this case we define the **weak derivative** $\partial^\nu u$ to be u_ν.

Proof: We sketch the main idea in the case $\Omega = \{x \in \mathbb{R}^n \mid x_1 > 0\}$. Assume that u has weak L^p derivatives up to order k. Let $\rho : \mathbb{R}^n \to \mathbb{R}$ be a smooth nonnegative function such that

$$\operatorname{supp} \rho \subset B_1, \qquad \int_{\mathbb{R}^n} \rho(x)\,dx = 1.$$

Define

$$\rho_\delta(x) = \delta^{-n}\rho(\delta^{-1}x).$$

Now choose a smooth cutoff function $\beta : \mathbb{R}^n \to [0,1]$ such that $\beta(x) = 1$ for $|x| < 1$ and $\beta(x) = 0$ for $|x| > 2$. Then the functions

$$u_\delta(x) = \beta(\delta x)u(x_1 + \delta, x_2, \ldots, x_n)$$

have compact support, are defined for $x_1 > -\delta$, and have weak L^p-derivatives up to order k. These converge in the L^p-norm to the weak derivatives of u. The functions

$$\rho_\delta * u_\delta(x) = \int_{\mathbb{R}^n} \rho_\delta(x - y)u_\delta(y)\,dy$$

are smooth with compact support and converge in the $W^{k,p}$-norm. The limit is u. Hence $u \in W^{k,p}(\Omega)$. The general case can be reduced to the above by choosing suitable local coordinates near every boundary point. $\qquad\Box$

It is somewhat less than obvious that a function $u \in W^{1,p}(\Omega)$ whose derivatives all vanish must be constant on every component of Ω. The proof requires the following fundamental estimate. As always, ∇u denotes the gradient of the function u. Recall that the mean value of a function is its integral over its domain of definition divided by the volume of the domain.

Lemma B.1.2 (Poincaré's inequality) *Let $1 < p < \infty$ and $\Omega \subset \mathbb{R}^n$ be a bounded open domain. Then there exists a constant $c = c(p, n, \Omega) > 0$ such that*

$$\|u\|_{L^p(\Omega)} \le c\,\|\nabla u\|_{L^p(\Omega)}$$

for every $u \in C_0^\infty(\Omega)$. If Ω is a square then this continues to hold for $u \in C^\infty(\bar{\Omega})$ with mean value zero.

Proof: The first statement is an easy exercise. The second statement on functions with mean value zero is proved by induction over n. Assume without loss of generality that Ω is the unit square $Q^n = \{x \in \mathbb{R}^n \mid 0 < x_j < 1\}$. For $n = 1$ the statement is again an easy exercise. Assume that the statement is proved for $n \ge 1$ and let $u \in C^\infty(Q^{n+1})$ be of mean value zero. Define $v \in C^\infty(Q^1)$ by

$$v(t) = \int_{Q^n} u(x_1, \ldots, x_n, t)\,dx_1 \cdots dx_n.$$

Since v is of mean value 0

$$\int_0^1 |v(t)|^p\,dt \le c_1 \int_0^1 |\dot{v}(t)|^p\,dt \le c_1 \int_{Q^{n+1}} \left|\frac{\partial u}{\partial x_{n+1}}\right|^p\,dx.$$

The last step follows from Hölder's inequality. By the induction hypothesis

$$\int_{Q^n} |u(x,t) - v(t)|^p \, dx \leq c_2 \int_{Q^n} |\nabla u(x,t)|^p \, dx.$$

Integrate over t and use the previous inequality to obtain the required estimate. $\square$

Corollary B.1.3 *Let $\Omega \subset \mathbb{R}^n$ be a bounded open domain and $u \in W^{1,p}(\Omega)$ with weak derivatives $\partial u / \partial x_j \equiv 0$ for $j = 1, \ldots, n$. Then u is constant on each connected component of Ω. If, moreover, $u \in W_0^{1,p}(\Omega)$ then $u \equiv 0$.*

Proof: First assume that Ω is a square and u has mean value zero. Approximate u in the $W^{1,p}$-norm by a sequence of smooth functions $u_\nu \in C^\infty(\bar{\Omega})$ with mean value zero. Then Poincaré's inequality shows that u_ν converges to zero in the L^p-norm. Hence $u = 0$. This shows u is constant on every square where its first derivatives vanish. Hence a function $u \in W^{1,p}(\Omega)$ with $\nabla u \equiv 0$ is constant on every connected component of Ω. The same argument works for $u \in W_0^{1,p}(\Omega)$. $\square$

A function with weak derivatives need not be continuous. Consider for example the function

$$u(x) = |x|^{-\alpha}$$

with $\alpha \in \mathbb{R}$ in the domain $\Omega = B_1 = \{x \in \mathbb{R}^n \,|\, |x| < 1\}$. Then

$$\frac{\partial u}{\partial x_j} = -\alpha x_j |x|^{-\alpha - 2}.$$

By induction,

$$|\partial^\nu u(x)| \leq c_\nu |x|^{-\alpha - |\nu|}.$$

Now the function $x \mapsto |x|^{-\beta}$ is integrable on B_1 if and only if $\beta < n$. Hence the derivatives of u up to order k will be p-integrable whenever

$$\alpha p + kp < n.$$

If $kp < n$ choose $0 < \alpha < n/p - k$ to obtain a function which is in $W^{k,p}(B_1)$ but not continuous at 0. For $kp > n$ this construction fails and, in fact, in this case every $W^{k,p}$-function is continuous. More precisely, define the **Hölder norm**

$$\|u\|_{C^\varepsilon} = \sup_{x,y \in \Omega} \frac{|u(x) - u(y)|}{|x - y|^\varepsilon} + \sup_{x \in \Omega} |u(x)|$$

for $0 < \varepsilon \leq 1$ and

$$\|u\|_{C^{k+\varepsilon}} = \sum_{|\nu| \leq k} \|\partial^\nu u\|_{C^\varepsilon}.$$

Denote by $C^{k+\varepsilon}(\Omega)$ the space of all C^k-functions $u : \Omega \to \mathbb{R}$ with finite Hölder norm $\|u\|_{C^{k+\varepsilon}}$.

Theorem B.1.4 *Assume $kp > n$. Then there exists a constant $c = c(k, p, \Omega) > 0$ such that*

$$\|u\|_{C^\varepsilon} \leq c \|u\|_{W^{k,p}}, \qquad \varepsilon = k - \frac{n}{p}.$$

In particular, by the Arzela-Ascoli theorem, the injection $W^{k,p}(\Omega) \hookrightarrow C^0(\Omega)$ is compact.

Theorem B.1.5 *Assume $kp < n$. Then there exists a constant $c = c(k, p, \Omega) > 0$ such that*

$$\|u\|_{L^q} \leq c \|u\|_{W^{k,p}}, \qquad q = \frac{np}{n - kp}.$$

If $q < np/(n - kp)$ then the injection $W^{k,p}(\Omega) \hookrightarrow L^q(\Omega)$ is compact.

These are the **Sobolev estimates**. The compactness statement in Theorem B.1.5 is known as **Rellich's theorem**. The case $kp = n$ is the borderline situation for these estimates. In this case the space $W^{k,p}$ does not embed into the space of continuous functions. Of particular interest here is the case where $n = p = 2$ and $k = 1$.

Remark B.1.6 In Lemma A.1.1 we have seen that there exists a sequence of function $u_j \in W^{1,2}(B_1)$ on the unit disc $B_1 = \{(x, y) \in \mathbb{R}^2 \mid x^2 + y^2 < 1\}$ such that

$$u_j(0) = 1, \qquad \lim_{j \to \infty} \|u_j\|_{W^{1,2}} = 0.$$

In 2 dimensions the relation between $W^{1,2}$ and C^0 is rather subtle. For example the function $u(e^{i\theta}) = \theta$, $0 \leq \theta < 2\pi$, (with a single jump discontinuity) does not extend to a $W^{1,2}$-function an B_1. On the other hand there exist discontinuous functions on S^1 which do extend to $W^{1,2}$-functions on B_1. $\qquad\qquad\square$

The case $kp > n$ should be viewed as the *good case* where *everything works*. For example composition with a smooth function and products.

Proposition B.1.7 *Assume $kp > n$. Then there exists a constant $c = c(k, p) > 0$ such that*

$$\|uv\|_{W^{k,p}} \leq c \left(\|u\|_{W^{k,p}} \|v\|_{L^\infty} + \|u\|_{L^\infty} \|v\|_{W^{k,p}} \right)$$

$$\|f \circ u\|_{W^{k,p}} \leq c \left(\|f\|_{C^k} + 1 \right) \|u\|_{W^{k,p}}$$

for $u, v \in C^\infty(\Omega)$ and $f \in C^k(\mathbb{R})$.

Proposition B.1.8 *Assume $kp > n$ and $f \in C^\infty(\mathbb{R})$. Then the map*

$$W^{k,p}(\Omega) \to W^{k,p}(\Omega) : u \mapsto f \circ u$$

is a C^∞-map of Banach spaces.

Remark B.1.9 Let M be an n-dimensional smooth compact manifold and $\pi : E \to M$ be a smooth vector bundle. A section $s : M \to E$ is said to be **of class $W^{k,p}$** if all its local coordinate representations are in $W^{k,p}$. This definition is independent of the choice of the coordinates. To see this note that if $\phi \in \text{Diff}(\mathbb{R}^n)$, $\Phi \in C^\infty(\mathbb{R}^n, \mathbb{R}^{N \times N})$ is matrix-valued function, and $u \in W^{k,p}_{\text{loc}}(\mathbb{R}^n, \mathbb{R}^N)$ then $\Phi(u \circ \phi) \in W^{k,p}(\Omega)$ for every bounded open set $\Omega \subset \mathbb{R}^n$ and there is an estimate

$$\|\Phi(u \circ \phi)\|_{W^{k,p}(\Omega)} \leq c \|u\|_{W^{k,p}(\Omega)}$$

with $c = c(\phi, \Phi, \Omega)$ independent of u. This holds even when $kp \leq n$. For functions with *values* in a manifold the situation is quite different. Proposition B.1.7 shows that for such functions it is required that $kp > n$. To define a norm on the space of $W^{k,p}$-sections one can take the sum of the $W^{k,p}$-norms over finitely many charts which cover M. $\qquad\qquad\square$

B.2 The Calderon-Zygmund inequality

Denote by

$$\Delta = \frac{\partial^2}{\partial x_1{}^2} + \frac{\partial^2}{\partial x_2{}^2} + \cdots + \frac{\partial^2}{\partial x_n{}^2}$$

the Laplace-operator on $\mathbb{R}^n$. A C^2-function $u : \Omega \to \mathbb{R}$ on an open set $\Omega \subset \mathbb{R}^n$ is called **harmonic** if $\Delta u = 0$. Harmonic functions are real analytic. (If $n = 2$ then a function is harmonic iff it is locally the real part of a holomorphic function.) Harmonic functions are characterized by the **mean value property** (see John [33])

$$u(x) = \frac{n}{\omega_n r^2} \int_{B_r(x)} u(\xi)\,d\xi, \qquad B_r(x) \subset \Omega.$$

Here $\omega_n = 2\pi^{n/2}\Gamma(n/2)^{-1}$ is the volume of the unit sphere in $\mathbb{R}^n$. In particular, $\omega_2 = 2\pi$.

The **fundamental solution** of Laplace's equation is

$$K(x) = \begin{cases} (2\pi)^{-1}\log|x|, & n = 2, \\ (2-n)^{-1}\omega_n^{-1}|x|^{2-n}, & n \geq 3. \end{cases}$$

Its first and second derivatives are given by

$$K_j(x) = \frac{x_j}{\omega_n|x|^n}, \qquad K_{jk}(x) = \frac{-nx_j x_k}{\omega_n|x|^{n+2}}, \qquad K_{jj}(x) = \frac{|x|^2 - nx_j^2}{\omega_n|x|^{n+2}}$$

where $K_j = \partial K/\partial x_j$ and $K_{jk} = \partial^2 K/\partial x_j \partial x_k$. In particular, $\Delta K = 0$. The function K and its first derivatives K_j are integrable on compact sets while the second derivatives are not. Hence $\partial_j(K * f) = K_j * f$ for compactly supported functions $f : \mathbb{R}^n \to \mathbb{R}$ but there is no such formula for the second derivatives. Moreover, since neither K nor its derivatives are integrable on $\mathbb{R}^n$, care must be taken for functions f which do not have compact support.

Every compactly supported C^2-function $u : \mathbb{R}^n \to \mathbb{R}$ satisfies

$$u = K * \Delta u, \qquad \partial_j u = K_j * \Delta u.$$

where $*$ denotes convolution. Conversely,

$$\Delta(K * f) = f, \qquad \Delta(K_j * f) = \partial_j f$$

for $f \in C_0^\infty(\mathbb{R}^n)$ (see [33]). This is **Poisson's identity**. In general $K * f$ will not have compact support. Since the second derivatives of K are not integrable on compact sets there exists a continuous function $f : \mathbb{R}^n \to \mathbb{R}$ such that $K * f \notin C^2$. For such a function f there is no classical solution of $\Delta u = f$. (This follows from the next two lemmata.) The situation is however quite different for weak solutions. Let $f \in L^1_{\mathrm{loc}}(\Omega)$. A function $u \in L^1_{\mathrm{loc}}(\Omega)$ is called a **weak solution** of $\Delta u = f$ if

$$\int_\Omega u(x)\Delta\phi(x)\,dx = \int_\Omega f(x)\phi(x)\,dx$$

for $\phi \in C_0^\infty(\Omega)$. Similarly $u \in L^1_{\mathrm{loc}}(\Omega)$ is called a weak solution of $\Delta u = \partial_j f$ with $f \in L^1_{\mathrm{loc}}$ if

$$\int_\Omega u(x)\Delta\phi(x)\,dx = -\int_\Omega f(x)\partial_j\phi(x)\,dx$$

for $\phi \in C_0^\infty(\Omega)$.

Lemma B.2.1 *Let $u, f \in L^1(\mathbb{R}^n)$ with compact support.*

(i) *u is a weak solution of $\Delta u = f$ if and only if $u = K * f$.*

(i) *u is a weak solution of $\Delta u = \partial_j f$ if and only if $u = K_j * f$.*

Proof: If $u = K * f$. Then

$$\int u \Delta \phi = \int (K * f) \Delta \phi = \int f(K * \Delta \phi) = \int f \phi$$

for $\phi \in C_0^\infty(\mathbb{R}^n)$. Conversely, suppose that u is a weak solution of $\Delta u = f$. Choose $\rho_\delta : \mathbb{R}^n \to \mathbb{R}$ as in the proof of Proposition B.1.1. Then

$$\int (\Delta \rho_\delta * u)\phi = \int u(\Delta \rho_\delta * \phi) = \int f(\rho_\delta * \phi) = \int (\rho_\delta * f)\phi.$$

for every $\phi \in C_0^\infty(\mathbb{R}^n)$ and hence $\Delta(\rho_\delta * u) = (\Delta \rho_\delta) * u = \rho_\delta * f$. This implies that $\rho_\delta * u - K * \rho_\delta * f$ is a bounded harmonic function converging to zero at ∞ and hence $\rho_\delta * u = K * \rho_\delta * f$. Take the limit $\delta \to 0$ to obtain $u = K * f$. This proves (i). The proof of (ii) is similar and is left to the reader. $\qquad \square$

Lemma B.2.2 (Weyl's lemma) *Every weak solution $u \in L_{\mathrm{loc}}^1(\Omega)$ of $\Delta u = 0$ is harmonic.*

Proof: Let ρ_δ be as in the proof of Lemma B.2.1. The function $u_\delta = \rho_\delta * u$ is harmonic in $\Omega_\delta = \{x \in \Omega \,|\, B_\delta(x) \subset \Omega\}$. Hence u_δ satisfies the mean value property. Since u_δ converges to u in the L^1-norm on every compact subset of Ω it follows that u has the mean value property. Hence u is harmonic (cf. [33]). $\qquad \square$

Theorem B.2.3 (Calderon-Zygmund inequality) *For $1 < p < \infty$ there exists a constant $c = c(n, p) > 0$ such that*

$$\|\nabla(K_j * f)\|_{L^p} \leq c \|f\|_{L^p} \tag{B.1}$$

for $f \in C_0^\infty(\mathbb{R}^n)$ and $j = 1, \ldots, n$.

This theorem is the fundamental estimate for the L^p-theory of elliptic operators. We include here a proof which is a modification of the one given by Gilbarg and Trudinger [23]. The proof requires the following four lemmata. The first is the case $p = 2$.

Lemma B.2.4 *The estimate (B.1) holds for $p = 2$.*

Proof: Since $u(x) = K_j * f(x)$ converges to zero as $|x|$ tends to infinity we have

$$\|\nabla u\|_{L^2}^2 = \langle \nabla u, \nabla u \rangle = -\langle u, \Delta u \rangle = -\langle u, \partial_j f \rangle = \langle \partial_j u, f \rangle \leq \|\nabla u\|_{L^2} \|f\|_{L^2}.$$

Here all inner products are L^2-inner products on $\mathbb{R}^n$. Divide both sides by $\|\nabla u\|_{L^2}$ to obtain the required estimate. $\qquad \square$

For any measurable function $f : \mathbb{R}^n \to \mathbb{R}$ define

$$\mu(t, f) = \left| \{ x \in \mathbb{R}^2 \,|\, |f(x)| > t \} \right|$$

for $t > 0$ where $|A|$ denotes the Lebesgues measure of the set A.

Lemma B.2.5 *For $1 \leq p < \infty$ and $f \in L^p(\mathbb{R}^n)$*

$$t^p \mu(t, f) \leq \int |f(x)|^p \, dx = p \int_0^\infty s^{p-1} \mu(s, f) \, ds.$$

Moreover,

$$\mu(t, f + g) \leq \mu(t/2, f) + \mu(t/2, g).$$

Proof: Integrate the function $F : \mathbb{R}^{n+1} \to \mathbb{R}$ defined by $F(x, t) = p t^{p-1}$ for $0 \leq t \leq |f(x)|$ and $F(x, t) = 0$ otherwise. $\qquad\square$

Apply the previous Lemma to the function $\partial_k(K_j * f)$. By Lemma B.2.4

$$\|\partial_k(K_j * f)\|_{L^2} \leq \|f\|_{L^2}$$

and hence

$$\mu(t, \partial_k(K_j * f)) \leq \frac{1}{t^2} \int |f(x)|^2 \, dx \tag{B.2}$$

The next two lemmata establish a similar inequality with the L^2-norm on the right replaced by the L^1-norm. Theorem B.2.3 is proved by interpolation for $1 < p < 2$.

Lemma B.2.6 *There exists a constant $c = c(n) > 0$ such that every function $f \in L^1(\mathbb{R}^n)$ with $\operatorname{supp} f \subset B_r$ and mean value 0 satisfies*

$$\mu(t, \partial_k(K_j * f)) \leq c \left(r^n + \frac{1}{t} \|f\|_{L^1} \right)$$

for $j, k = 1, \ldots, n$.

Proof: Denote $u_{kj} = \partial_k(K_j * f)$. For $|x| > r$

$$
\begin{aligned}
u_{kj}(x) &= \int_{B_r} (\partial_k K_j(x - y) - \partial_k K_j(x)) f(y) \, dy \\
&\leq \max_{y \in B_r} |\partial_k K_j(x - y) - \partial_k K_j(x)| \, \|f\|_{L^1} \\
&\leq r \max_{y \in B_r} |\nabla \partial_k K_j(x - y)| \, \|f\|_{L^1} \\
&\leq r \max_{y \in B_r} \frac{n^{3/2}(n + 2)}{\omega_n |x - y|^{n+1}} \|f\|_{L^1} \\
&\leq \frac{n^{3/2}(n + 2)r}{\omega_n(|x| - r)^{n+1}} \|f\|_{L^1}.
\end{aligned}
$$

Hence

$$
\begin{aligned}
\int_{|x| > 2r} |u_{kj}(x)| \, dx &\leq \frac{n^{3/2}(n + 2)r}{\omega_n} \int_{|x| > 2r} \frac{dx}{(|x| - r)^{n+1}} \|f\|_{L^1} \\
&= \frac{n^{3/2}(n + 2)r}{\omega_n} \int_{2r}^\infty \frac{\omega_n \rho^{n-1} \, d\rho}{(\rho - r)^{n+1}} \|f\|_{L^1} \\
&\leq 2^{n-1} n^{3/2}(n + 2)r \int_{2r}^\infty \frac{d\rho}{(\rho - r)^2} \|f\|_{L^1} \\
&= 2^{n-2} n^{3/2}(n + 2) \|f\|_{L^1}.
\end{aligned}
$$

This implies

$$
\begin{aligned}
t\mu(t, u_{kj}) &\leq t|B_{2r}| + |\{x \in \mathbb{R}^n \,|\, |u_{kj}(x)| > t,\ |x| > 2r\}| \\
&\leq t|B_{2r}| + \int_{|x|>2r} |u_{kj}(x)|\, dx \\
&= \frac{t\omega_n 2^n r^n}{n} + 2^{n-2}n^{3/2}(n+2)\,\|f\|_{L^1}.
\end{aligned}
$$

$\square$

Lemma B.2.7 *There exists a constant $c = c(n) > 0$ such that the characteristic function $f = \chi_Q$ of any square $Q \subset \mathbb{R}^n$ satisfies*

$$
\mu(t, \partial_k(K_j * f)) \leq \frac{c}{t} \int |f(x)|\, dx
$$

for $j, k = 1, \ldots, n$.

Proof: Assume without loss of generality that $Q = Q_r = \{x \in \mathbb{R}^n \,|\, |x_j| < r\}$. For $t > 1$ the left hand side is zero. Hence assume $0 < t \leq 1$ and write

$$
f = f_0 + f_1, \qquad f_0 = t\chi_{Q_R}, \qquad f_1 = \chi_{Q_r} - t\chi_{Q_R}, \qquad R = t^{-1/n}r.
$$

Then

$$
\int f_0 = \int f = (2r)^n, \qquad \operatorname{supp} f_1 \subset B_{\sqrt{n}R},
$$

and f_1 is of mean value zero. Let $c_1 > 0$ be the constant of Lemma B.2.6. Then

$$
\mu(t, \partial_k(K_j * f_1)) \leq c_1 \left(\frac{n^{n/2}r^n}{t} + \frac{1}{t}\|f_1\|_{L^1} \right) \leq \frac{c_2}{t}\|f\|_{L^1}.
$$

where $c_2 = c_1(n^{n/2}2^{-n} + 2)$. By Lemma B.2.5 and Lemma B.2.4

$$
\mu(t, \partial_k(K_j * f_0)) \leq \frac{1}{t^2}\|\partial_k(K_j * f_0)\|_{L^2} \leq \frac{1}{t^2}\|f_0\|_{L^2} = \frac{1}{t}\|f_0\|_{L^1} = \frac{1}{t}\|f\|_{L^1}.
$$

Again by Lemma B.2.5

$$
\mu(t, \partial_k(K_j * f)) \leq \mu(t/2, \partial_k(K_j * f_0)) + \mu(t/2, \partial_k(K_j * f_1)) \leq \frac{2 + 2c_2}{t}\|f\|_{L^1}. \quad \square
$$

The previous estimate holds for all $f \in L^2 \cap L^1$ [23]. However, for the proof of Theorem B.2.3 it is only needed in the case where f is a step function.

Proof of Theorem B.2.3: First assume $1 < p < 2$. It suffices to prove the estimate for characteristic functions of squares. Hence assume $f = \chi_Q$ for some open square $Q \subset \mathbb{R}^n$. Then by Lemma B.2.5

$$
\begin{aligned}
\int_{\mathbb{R}^n} |\partial_k(K_j * f)(x)|^p\, dx &= p\int_0^\infty t^{p-1}\mu(t, \partial_k(K_j * f))\, dt \\
&\leq \left(pc\int_0^1 t^{p-2}\, dt + p\int_1^\infty t^{p-3}\, dt \right)\|f\|_{L^1} \\
&= \left(\frac{pc}{p-1} + \frac{p}{2-p} \right)\|f\|_{L^p}.
\end{aligned}
$$

Here we have used Lemma B.2.7 for $t < 1$ and (B.2) for $t > 1$. This proves the estimate for $1 < p < 2$. For $2 < p < \infty$ we use duality. Let $1 < q < 2$ such that $1/p + 1/q = 1$. Then

$$\int g(x)\partial_k(K_j * f)(x)\,dx = \int \partial_k(K_j * g)(x)f(x) \le c\,\|f\|_{L^p}\,\|g\|_{L^q}$$

and hence $\|\partial_k(K_j * f)\|_{L^p} \le c\|f\|_{L^p}$. $\qquad\square$

Theorem B.2.8 (Local regularity) *Let $1 < p < \infty$, $k \ge 0$ be an integer, and $\Omega \subset \mathbb{R}^n$ be an open domain. If $u \in L^p_{\mathrm{loc}}(\Omega)$ is a weak solution of $\Delta u = f$ with $f \in W^{k,p}_{\mathrm{loc}}(\Omega)$ then $u \in W^{k+2,p}_{\mathrm{loc}}(\Omega)$. Moreover, for every compact set $Q \subset \Omega$ there exists a constant $c = c(k,p,n,Q,\Omega) > 0$ such that*

$$\|u\|_{W^{k+2,p}(Q)} \le c\left(\|\Delta u\|_{W^{k,p}(\Omega)} + \|u\|_{L^p(\Omega)}\right)$$

for $u \in C^\infty(\bar{\Omega})$.

Proof: Let u and f be as in (i) and Q as in (ii). Choose an open neighborhood U of Q such that $\mathrm{cl}(U) \subset \Omega$. Let $\beta \in C^\infty_0(\Omega)$ be a smooth cutoff function such that $\beta(x) = 1$ for $x \in U$ and define

$$v = K * \beta f.$$

It follows from Theorem B.2.3 that $v \in W^{k+2,p}_{\mathrm{loc}}(\Omega)$. By Lemma B.2.1 v is a weak solution of $\Delta v = \beta f$. Hence the restriction of $u - v$ to U is a weak solution of $\Delta(u - v) = 0$. By Weyl's lemma $u - v$ is real analytic in U. Hence $u \in W^{k+2,p}(Q)$. This proves the first statement.

By Theorem B.2.3, the function $v = K * \beta\Delta u$ satisfies an estimate

$$\|v\|_{W^{k+2,p}(U)} \le c_1 \|\beta\Delta u\|_{W^{k,p}(\Omega)} \le c_2 \|\Delta u\|_{W^{k,p}(\Omega)}.$$

The function $v - u$ is harmonic in U. By the mean value property for harmonic functions there exists a constant $c_3 > 0$ such that

$$\|v - u\|_{W^{k+2,p}(Q)} \le c_3 \|v - u\|_{L^p(U)} \le c_3\left(\|v\|_{W^{k+2,p}(U)} + \|u\|_{L^p(U)}\right).$$

Take the sum of these inequalities to obtain the required estimate. $\qquad\square$

Sometimes it is useful to consider weak solutions of $\Delta u = f$ where f is not a function but a distribution in $W^{-1,p}$. We rephrase this in terms of weak solutions of $\Delta u = \mathrm{div} f$ with $f \in L^p$.

Theorem B.2.9 *Let $1 < p < \infty$, $k \ge 0$ be an integer, and $\Omega \subset \mathbb{R}^n$ be an open domain. Assume that $u \in L^p_{\mathrm{loc}}(\Omega)$ and $f = (f_0, \dots, f_n) \in L^p_{\mathrm{loc}}(\Omega, \mathbb{R}^{n+1})$ satisfy*

$$\int_\Omega u(x)\Delta\phi(x)\,dx = \int_\Omega f_0(x)\phi(x)\,dx - \sum_{j=1}^n \int_\Omega f_j(x)\partial_j\phi(x)\,dx$$

for every $\phi \in C^\infty_0(\Omega)$. Then $u \in W^{1,p}_{\mathrm{loc}}(\Omega)$ and for every compact set $Q \subset \Omega$ there is an estimate

$$\|u\|_{W^{1,p}(Q)} \le c\left(\|f\|_{L^p(\Omega)} + \|u\|_{L^p(\Omega)}\right)$$

where $c = c(p,n,Q,\Omega) > 0$ is independent of u.

Proof: Choose an open neighbourhood U of Q with $\text{cl}(U) \subset \Omega$. Let $\beta \in C_0^\infty(\Omega)$ be a smooth cutoff function such that $\beta(x) = 1$ for $x \in U$. Define

$$v = K * \beta f_0 + \sum_{j=1}^n K_j * \beta f_j.$$

It follows from Theorem B.2.3 that $v \in W_{\text{loc}}^{1,p}(\Omega)$ and there an estimate

$$\|v\|_{W^{1,p}(U)} \leq c_1 \|\beta f\|_{L^p(\Omega)} \leq c_1 \|f\|_{L^p(\Omega)}.$$

Here we have also used Poincaré's inequality. Now, by Lemma B.2.1, v is a weak solution of

$$\Delta v = \beta f_0 + \sum_{j=1}^n \partial_j(\beta f_j)$$

where $\partial_j(\beta f_j)$ is to be understood as a distribution. Hence the restriction of $u - v$ to Q is a weak solution of $\Delta(u - v) = 0$. By Weyl's lemma $u - v$ is harmonic in U. Hence $u \in W^{1,p}(Q)$. Moreover, by the mean value property for harmonic functions, there exists a constant $c_2 > 0$ such that

$$\|v - u\|_{W^{1,p}(Q)} \leq c_2 \|v - u\|_{L^p(U)} \leq c_2 \left(\|v\|_{W^{1,p}(U)} + \|u\|_{L^p(U)} \right).$$

Take the sum of these inequalities to obtain the reqired estimate. $\qquad\square$

B.3 Cauchy-Riemann operators

Identify $\mathbb{R}^2 = \mathbb{C}$ with $z = x + iy$. For $z = x + iy$ and $\zeta = \xi + i\eta$ denote

$$\langle z, \zeta \rangle = x\xi + y\eta.$$

In this section all functions take values in $\mathbb{C}$. For example $C_0^\infty(\Omega)$ denotes the space of smooth compactly supported functions $\Omega \to \mathbb{C}$. Consider the first order differential operators

$$\frac{\partial}{\partial \bar{z}} = \frac{1}{2}\left(\frac{\partial}{\partial x} + i\frac{\partial}{\partial y}\right), \qquad \frac{\partial}{\partial z} = \frac{1}{2}\left(\frac{\partial}{\partial x} - i\frac{\partial}{\partial y}\right).$$

A C^1-function $u : \Omega \to \mathbb{C}$ on an open set $\Omega \subset \mathbb{C}$ satisfies the **Cauchy-Riemann equation** $\partial u/\partial \bar{z} = 0$ if and only if it is holomorphic. It satisfies $\partial u/\partial z = 0$ iff it is anti-holomorphic. The following lemma shows that the function

$$N(z) = \frac{1}{\pi z}.$$

is the fundamental solution of the Cauchy-Riemann operator $\partial/\partial \bar{z}$.

Lemma B.3.1 *If $f : \mathbb{C} \to \mathbb{C}$ is a smooth compactly supported function then $u = N * f$ is a solution of the inhomogeneous Cauchy Riemann equations $\partial u/\partial \bar{z} = f$.*

Proof: Note that

$$\Delta = 4\frac{\partial}{\partial z}\frac{\partial}{\partial \bar{z}}, \qquad N = 4\frac{\partial K}{\partial z}, \qquad K(z) = \frac{\log|z|}{2\pi}.$$

Hence

$$\frac{\partial}{\partial \bar{z}}(N * f) = 4\frac{\partial}{\partial \bar{z}}\left(\frac{\partial K}{\partial z} * f\right) = 4\frac{\partial}{\partial \bar{z}}\frac{\partial}{\partial z}(K * f) = \Delta(K * f) = f$$

for $f \in C_0^\infty(\mathbb{C})$ $\qquad\qquad\square$

By the divergence theorem

$$\int_\Omega \langle \partial v/\partial z, u\rangle + \int_\Omega \langle v, \partial u/\partial \bar{z}\rangle = 0$$

for $u, v \in C_0^\infty(\Omega)$. In other words $-\partial/\partial z$ is the formal adjoint operator of $\partial/\partial \bar{z}$. A function $u \in L_{\mathrm{loc}}^1(\Omega)$ is called a **weak solution** of $\partial u/\partial \bar{z} = f$ for $f \in L_{\mathrm{loc}}^1(\Omega)$ if

$$\int_\Omega \langle \partial \phi/\partial z, u\rangle + \int_\Omega \langle \phi, f\rangle = 0 \qquad\qquad (\text{B.3})$$

for $\phi \in C_0^\infty(\Omega)$.

Lemma B.3.2 *Let $u, f \in L^p(\mathbb{C})$ with compact support. Then u is a weak solution of $\partial u/\partial \bar{z} = f$ if and only if $u = N * f$.*

Proof: Assume first $u = N * f$. Then

$$\int_\Omega \langle u, \partial \phi/\partial z\rangle = \int_\Omega \langle N * f, \partial \phi/\partial z\rangle = -\int_\Omega \langle f, \bar{N} * \partial \phi/\partial z\rangle = -\int_\Omega \langle f, \phi\rangle$$

for $\phi \in C_0^\infty(\Omega)$. Hence u is a weak solution of $\partial u/\partial \bar{z} = f$. Conversely, suppose that u satisfies (B.3) and choose $\rho_\delta : \mathbb{C} \to \mathbb{R}$ as in the proof of Proposition B.1.1 with $n = 2$. Then

$$\frac{\partial}{\partial \bar{z}}\rho_\delta * u = \rho_\delta * f.$$

Since $\rho_\delta * u \in C_0^\infty(\mathbb{C})$ it follows from by Lemma B.3.1 that $\rho_\delta * u - N * \rho_\delta * f$ is a bounded holomorphic function converging to zero at ∞ and hence $\rho_\delta * u = N * \rho_\delta * f$. Take the limit $\delta \to 0$ to obtain $u = N * f$. $\qquad\square$

Lemma B.3.3 *Every weak solution $u \in L_{\mathrm{loc}}^1(\Omega)$ of $\partial u/\partial \bar{z} = 0$ is holomorphic.*

Proof: If u is a weak soluion of $\partial u/\partial \bar{z} = 0$ then its real and imaginary part are weak solutions of Laplace's equation. By Weyl's lemma they are smooth. Hence u is holomorphic. $\qquad\square$

Theorem B.3.4 (Local regularity) *Let $1 < p < \infty$, $k \geq 0$ be an integer, and $\Omega \subset \mathbb{C}$ be an open domain. If $u \in L^p_{\text{loc}}(\Omega)$ is a weak solution of $\partial u/\partial \bar{z} = f$ with $f \in W^{k,p}_{\text{loc}}(\Omega)$ then $u \in W^{k+1,p}_{\text{loc}}(\Omega)$. Moreover, for every compact set $Q \subset \Omega$ there exists a constant $c = c(k,p,Q,\Omega) > 0$ such that*

$$\|u\|_{W^{k+1,p}(Q)} \leq c \left(\left\| \frac{\partial u}{\partial \bar{z}} \right\|_{W^{k,p}(\Omega)} + \|u\|_{L^p(\Omega)} \right)$$

for $u \in C^\infty(\bar{\Omega})$.

Proof: Let u and f be as in (i) and Q as in (ii). Choose a compact neighborhood U of Q such that $U \subset \Omega$. Let $\beta \in C^\infty_0(\Omega)$ be a smooth cutoff function such that $\beta(x) = 1$ for $x \in U$ and define

$$v = N * \beta f.$$

By Theorem B.2.3 $v \in W^{k+1,p}_{\text{loc}}(\Omega)$. By Lemma B.3.2 v is a weak solution of $\partial v/\partial \bar{z} = \beta f$. Hence the restriction of $u - v$ to U is a weak solution of

$$\frac{\partial(u - v)}{\partial \bar{z}} = 0.$$

By Lemma B.3.3 $u - v$ is holomorphic in $\text{int}(U)$. Hence $u \in W^{k+1,p}(Q)$. This proves the first statement. The proof of the estimate is as in Theorem B.2.8 and is left to the reader. $\square$

Exercise B.3.5 Use the previous theorem to prove the estimate (3.3) of Chapter 3. Use it also to prove that the cokernel of D_u is the kernel of D^*_u, or more precisely, if $\eta \in L^q(\Lambda^{0,1}T^*\Sigma \otimes_J u^*TM)$ satisfies $\langle \eta, D_u\xi \rangle = 0$ for all $\xi \in W^{1,p}(u^*TM)$ then $\eta \in W^{1,q}$ and $D^*_u\eta = 0$. **Hint:** In a suitable trivialization u^*TM the operator D_u is of the form $\partial/\partial \bar{z} +$ zero-th order term. Use the formula of Remark 3.3.3. $\square$

B.4 Elliptic bootstrapping

In this section we shall prove the following two theorems about the smoothness of J-holomorphic curves and compactness for sequences with uniform bounds on the L^p-norm of the derivatives with $p > 2$. Assume that (M, J) is an almost complex manifold. Let Σ be an oriented Riemann surface without boundary with complex structure j.

Theorem B.4.1 (Regularity) *If $u : \Sigma \to M$ is a J-holomorphic curve of class $W^{k,p}$ with $kp > 2$ then u is smooth.*

Theorem B.4.2 (Compactness) *Let J_ν be a sequence of almost complex structures on M converging to J in the C^∞-topology and j_ν be a sequence of complex structures on Σ converging to j in the C^∞-topology. Let $U_\nu \subset \Sigma$ be an increasing sequence of open sets whose union is Σ and $u_\nu : U_\nu \to M$ be a sequence of (j_ν, J_ν)-holomorphic curves. Assume that for every compact set $Q \subset \Sigma$ there exists a compact set $K \subset M$ and constants $p > 2$ and $c > 0$ such that*

$$\|du_\nu\|_{L^p(Q)} \leq c, \qquad u_\nu(Q) \subset K$$

for ν sufficiently large. Then a subsequence of the u_ν converges uniformly with all derivatives on compact sets to a (j, J)-holomorphic curve $u : \Sigma \to M$.

Both these theorems are obvious when J is integrable and Σ is closed because each component of a J-holomorphic curve in holomorphic coordinates is a harmonic function. In particular, the compactness theorem follows from the mean value property of harmonic functions. The general case is considerably harder.

We begin by proving a local estimate for the solutions of linear Cauchy-Riemann equations. Let $\Omega \subset \mathbb{C}$ be an open set and $u : \Omega \to \mathbb{R}^{2n}$ be a solution of the linear PDE

$$\partial_s u + J \partial_t u = \eta, \tag{B.4}$$

where $J : \Omega \to \mathbb{R}^{2n \times 2n}$ with $J^2 = -\mathbb{1}$ and $\eta : \Omega \to \mathbb{R}^{2n}$. We shall first prove regularity of u under the weakest possible regularity assumptions on the almost complex structure J and the function η. Eventually we shall consider the semilinear case where J is replaced by a function of the form $J \circ u$. Here J is a smooth almost complex structure on $\mathbb{R}^{2n}$.

Denote by $W^{k,p}_{\text{loc}}(\Omega, \mathbb{R}^{2n})$ the set of those functions $u : \Omega \to \mathbb{R}^{2n}$ such that the restriction of u to any compact subset of Ω is of class $W^{k,p}$. We shall first assume that the almost complex structure J is of class $W^{1,p}_{\text{loc}}$. If u is a solution of (B.4) then, by partial integration,

$$\int_\Omega \langle \partial_s \phi + J^T \partial_t \phi, u \rangle \, ds\, dt = - \int_\Omega \langle \phi, \eta + (\partial_t J) u \rangle \, ds\, dt \tag{B.5}$$

for every test function $\phi \in C_0^\infty(\Omega, \mathbb{R}^{2n})$. (Here J^T denotes the transpose of J.) The next lemma asserts the converse.

Lemma B.4.3 *Assume $J \in W^{1,p}_{\text{loc}}$ and $\eta \in L^{pq/(p+q)}_{\text{loc}}$ where $p > 2$ and $p/(p-1) < q \leq \infty$. If $u \in L^q_{\text{loc}}$ satisfies (B.5) for every $\phi \in C_0^\infty$ then $u \in W^{1,pq/(p+q)}_{\text{loc}}$ and u satisfies (B.4) almost everywhere. Moreover, for every compact set $Q \subset \Omega$ there is an estimate*

$$\|u\|_{W^{1,pq/(p+q)}(Q)} \leq c \left(\|u\|_{L^q(\Omega)} + \|\eta\|_{L^{pq/(p+q)}(\Omega)} \right)$$

where the constant $c = c(p, q, Q, \Omega, \|J\|_{W^{1,p}}) > 0$ is independent of u. [1]

Proof: Let $\psi : \Omega \to \mathbb{R}^{2n}$ be a smooth test function with compact support and define $\phi = \partial_s \psi - J^T \partial_t \psi$. Then

$$\partial_s \phi + J^T \partial_t \phi = \Delta \psi - (\partial_s J)^T \partial_t \psi + (\partial_t J)^T J^T \partial_t \psi$$

and hence (B.5) implies

$$\int_\Omega \langle \Delta \psi, u \rangle = - \int_\Omega \langle \partial_s \psi, f \rangle + \int_\Omega \langle \partial_t \psi, g \rangle. \tag{B.6}$$

where

$$f = (\partial_t J) u + \eta, \qquad g = (\partial_s J) u + J \eta. \tag{B.7}$$

---

[1] In the case $q = \infty$ we have $\eta \in L^p_{\text{loc}}$ and $u \in W^{1,p}_{\text{loc}}$ and the estimate holds with the corresponding norms.

By assumption and Hölder's inequality the functions f and g are of class $L^{pq/(p+q)}_{\mathrm{loc}}$. Equation B.6 asserts that u is a weak solution of

$$\Delta u = \partial_s f - \partial_t g$$

where the right hand side is to be understood in the distributional sense. By Theorem B.2.9 we have $u \in W^{1,pq/(p+q)}_{\mathrm{loc}}$ and for every compact set $Q \subset \Omega$ there is an estimate

$$
\begin{aligned}
\|u\|_{W^{1,pq/(p+q)}(Q)} \ &\leq\ c_1 \left(\|f\|_{L^{pq/(p+q)}(\Omega)} + \|g\|_{L^{pq/(p+q)}(\Omega)} + \|u\|_{L^{pq/(p+q)}(\Omega)} \right) \\
&\leq\ c_1 \left(\|J\|_{W^{1,p}(\Omega)} \|u\|_{L^q(\Omega)} + \|u\|_{L^{pq/(p+q)}(\Omega)} \right. \\
&\qquad \left. + \|J\|_{L^\infty(\Omega)} \|\eta\|_{L^{pq/(p+q)}(\Omega)} + \|\eta\|_{L^{pq/(p+q)}(\Omega)} \right) \\
&\leq\ c_2 \left(\|u\|_{L^q(\Omega)} + \|\eta\|_{L^{pq/(p+q)}(\Omega)} \right).
\end{aligned}
$$

Here the constants c_1 and c_2 depend on p, q, Q, Ω, and the $W^{1,p}$-norm of J but not on u. If follows from integration by parts that u satisfies (B.4) almost everywhere. $\square$

Lemma B.4.4 *Given $p > 2$ there exists a finite increasing sequence $q_0 < q_1 < \cdots < q_\ell$ such that*

$$\frac{p}{p-1} < q_0 \leq p, \qquad q_{\ell-1} < \frac{2p}{p-2} < q_\ell,$$

and

$$q_{j+1} = \frac{2r_j}{2 - r_j}, \qquad r_j = \frac{pq_j}{p + q_j},$$

for $j = 0, \ldots, \ell - 1$.

Proof: Consider the map $h : (p/(p-1), 2p/(p-2)) \to (2, \infty)$ defined by

$$h(q) = \frac{2pq}{2p + 2q - pq} = \frac{2r}{2 - r}, \qquad r = \frac{pq}{p+q} < 2.$$

The condition $r < 2$ is equivalent to $q < 2p/(p-2)$. The map h is a monotonically increasing diffeomorphism such that $h(q) > q$. Now choose the sequence q_j such that $q_{j+1} = h(q_j)$. $\square$

Lemma B.4.5 *Assume $p > 2$ and $1 < r \leq p$. If $f \in W^{1,p}$ and $g \in W^{1,r}$ then $fg \in W^{1,r}$ with*

$$\|fg\|_{W^{1,r}} \leq c \|f\|_{W^{1,p}} \|g\|_{W^{1,r}}.$$

Proof: Examine the L^r-norm of the expression expression $d(fg) = (df)g + f(dg)$. The term $f(dg)$ can be estimated by the sup-norm of f and the $W^{1,r}$-norm of g. The term $(df)g$ can be estimated by

$$\|(df)g\|_{L^r} \leq \|df\|_{L^p} \|g\|_{L^q}, \qquad \frac{1}{p} + \frac{1}{q} = \frac{1}{r}.$$

If $r < 2$ then, since $p > 2$, we have $q = pr/(p-r) < 2r/(2-r)$ and the L^q norm of g can be estimated by the $W^{1,r}$-norm. The latter is obvious when $r \geq 2$. $\square$

Lemma B.4.6 *Assume* $J \in W^{1,p}_{\text{loc}}$ *and* $\eta \in L^p_{\text{loc}}$ *where* $p > 2$. *If* $u \in L^p_{\text{loc}}$ *satisfies (B.5) for every* $\phi \in C^\infty_0$ *then* $u \in W^{1,p}_{\text{loc}}$. *Moreover, for every compact set* $Q \subset \Omega$ *there is an estimate*

$$\|u\|_{W^{1,p}(Q)} \leq c \left(\|u\|_{L^p(\Omega)} + \|\eta\|_{L^p(\Omega)} \right)$$

with $c = c(p, Q, \Omega, \|J\|_{W^{1,p}}) > 0$ *independent of* u.

Proof: By Lemma B.4.3 and the Sobolev embedding theorem we have

$$u \in L^q_{\text{loc}}, \quad \frac{p}{p-1} < q < \frac{2p}{p-2} \quad \Longrightarrow \quad u \in W^{1,r} \subset L^{q'}_{\text{loc}},$$

where $r = pq/(p+q) < 2$ and $q' = 2r/(2-r) > q$. Now choose q_j and r_j as in Lemma B.4.4. By induction we have $u \in W^{1,r_j}_{\text{loc}}$ for every j and

$$
\begin{aligned}
\|u\|_{W^{1,r_j}(Q_j)} &\leq c_1 \left(\|u\|_{L^{q_j}(Q_{j-1})} + \|\eta\|_{L^{r_j}(Q_{j-1})} \right) \\
&\leq c_2 \left(\|u\|_{W^{1,r_{j-1}}(Q_{j-1})} + \|\eta\|_{L^p(\Omega)} \right) \\
&\leq c_3 \left(\|u\|_{W^{1,r_0}(Q_0)} + \|\eta\|_{L^p(\Omega)} \right) \\
&\leq c_4 \left(\|u\|_{L^p(\Omega)} + \|\eta\|_{L^p(\Omega)} \right)
\end{aligned}
$$

where $Q_{j+1} \subset \text{int}(Q_j)$ and $Q \subset \text{int}(Q_\ell)$. In the first and last inequality we have used Lemma B.4.3. In the last inequality we have also used $q_0 \leq p$. The second inequality follows from the Sobolev embedding theorem and the third by induction. With $j = \ell$ we have $r_\ell = pq_\ell/(p+q_\ell) > 2$ and hence u is continuous. Now use Lemma B.4.3 again with $q = \infty$. $\square$

Proposition B.4.7 *Assume* $J \in W^{k,p}_{\text{loc}}$ *and* $\eta \in W^{k,p}_{\text{loc}}$ *where* $p > 2$ *and* $k \geq 1$. *If* $u \in L^p_{\text{loc}}$ *satisfies (B.5) for every* $\phi \in C^\infty_0$ *then* $u \in W^{k+1,p}_{\text{loc}}$. *Moreover, for every compact set* $Q \subset \Omega$ *there is an estimate*

$$\|u\|_{W^{k+1,p}(Q)} \leq c \left(\|u\|_{W^{k,p}(\Omega)} + \|\eta\|_{W^{k,p}(\Omega)} \right)$$

with $c = c(p, Q, \Omega, \|J\|_{W^{k,p}}) > 0$ *independent of* u.

Proof: First assume $k = 1$. Then

$$u \in W^{1,q}_{\text{loc}}, \quad \frac{p}{p-1} < q \leq \infty \quad \Longrightarrow \quad u \in W^{2,pq/(p+q)}_{\text{loc}} \qquad \text{(B.8)}$$

and u satisfies the obvious estimate. To see this note that the function

$$u' = \partial_s u \in L^q_{\text{loc}}$$

is a weak solution of (B.4) with η replaced by

$$\eta' = \partial_s \eta - (\partial_s J) \partial_t u \in L^{pq/(p+q)}_{\text{loc}}.$$

Hence it follows from Lemma B.4.3 that $u' \in W_{\text{loc}}^{1,pq/(p+q)}$. By Lemma B.4.5 $\partial_t u = J(\partial_s u - \eta) \in W_{\text{loc}}^{1,pq/(p+q)}$ and hence $u \in W_{\text{loc}}^{2,pq/(p+q)}$.

To prove the statement for $k = 1$ choose sequences q_j and r_j as in Lemma B.4.4 and use (B.8) inductively as in the proof of Lemma B.4.6 to obtain $u \in W_{\text{loc}}^{2,r_j}$ for every j. With $j = \ell$ it follows that u is continuously differentiable and, by (B.8) with $q = \infty$, we have $u \in W_{\text{loc}}^{2,p}$. This proves the assertion for $k = 1$. For general k it is proved by induction. Assume the proposition is true for $k \geq 1$. Suppose that $J \in W_{\text{loc}}^{k+1,p}$ and $\eta \in W_{\text{loc}}^{k+1,p}$. Apply the induction hypothesis to u' and η' as above to obtain that $\partial_s u = u'$ and $\partial_t u = J(\partial_s u - \eta)$ are of class $W_{\text{loc}}^{k+1,p}$. Hence $u \in W_{\text{loc}}^{k+2,p}$ and this proves the proposition with k replaced by $k + 1$. $\square$

Proof of Theorem B.4.1: It suffices to prove the theorem in local coordinates. Hence we shall assume that $\Omega \subset \mathbb{C}$ is an open set and $u : \Omega \to \mathbb{R}^{2n}$ is a solution of the PDE

$$\frac{\partial u}{\partial s} + J(u)\frac{\partial u}{\partial t} = 0,$$

where $J : \mathbb{R}^{2n} \to \mathbb{R}^{2n \times 2n}$ is a smooth almost complex structure. Now assume first that $k = 1$. Then $u \in W_{\text{loc}}^{1,p}(\Omega, \mathbb{R}^{2n})$ satisfies the requirements of Proposition B.4.7 with $k = 1$, $\eta = 0$, and J replaced by $J \circ u \in W_{\text{loc}}^{1,p}$. Hence $u \in W_{\text{loc}}^{2,p}$ and $J \circ u \in W_{\text{loc}}^{2,p}$. Continue by induction to obtain that $u \in W_{\text{loc}}^{k,p}$ for every k. Hence u is smooth. This proves the theorem for $k = 1$. If $u \in W_{\text{loc}}^{k,p}$ with $(k-1)p \geq 2$ then $u \in W_{\text{loc}}^{1,q}$ for every q and the statement follows from the case $k = 1$. If $(k-1)p < 2$ then $u \in W_{\text{loc}}^{1,q}$ with $q = 2p/(2 - kp + p) > 2$ and the statement follows again from the case $k = 1$. $\square$

Proof of Theorem B.4.2: Since the inclusion $W^{1,p} \hookrightarrow C$ is compact for $p > 2$ we may assume that u_ν converges uniformly to a continuous function $u : \Sigma \to M$. Now, in local coordinates, we may assume that $u_\nu : \Omega \to \mathbb{R}^{2n}$ is a sequence of smooth solutions of the PDE

$$\partial_s u_\nu + J_\nu(u_\nu)\partial_t u_\nu = 0,$$

where J_ν converges to J in the C^∞-topology. It follows from Proposition B.4.7 by induction that

$$\sup_\nu \|u\|_{W^{k,p}(Q)} < \infty$$

for every k and every compact subset $Q \subset \Omega$. By the Arzela-Ascoli theorem the inclusion $W^{k,p} \hookrightarrow C^{k-1}$ is compact. This proves the existence of a subsequence which converges in the C^{k-1}-topology on every compact subset of Σ. The theorem follows by choosing a diagonal subsequence. $\square$

Exercise B.4.8 Find the simplest proof you can of Theorems B.4.1 and B.4.2 when J is integrable and Σ is closed. $\square$

Bibliography

[1] N. Aronszajn, A unique continuation theorem for elliptic differential equations or inequalities of the second order, *J. Math. Pures Appl.* **36** (1957), 235–239.

[2] P.S. Aspinwall and D.R. Morrison, Topological Field Theory and rational curves, *Commun. Math. Phys.* **151** (1993), 245–262.

[3] A. Astashkevich and V. Sadov, Quantum cohomology of partial flag manifolds $F_{n_1,\dots,n_k}$, Preprint (1993).

[4] M.F. Atiyah, New invariants for three and four dimensional manifolds, *Proc. Symp. Pure Math.* **48** (1988), 285–299.

[5] M. Audin and F. Lafontaine ed., *Holomorphic Curves in Symplectic Geometry*, Progress in Math **117**, Birkhauser (1994).

[6] A. Bertram, G. Daskalopoulos, and R. Wentworth, Gromov invariant for holomorphic maps from Riemann surfaces to Grassmannians, Preprint, (1993).

[7] M. Callahan, PhD thesis, Oxford (1994).

[8] P. Candelas and X.C. de la Ossa, Moduli space of Calabi-Yau manifolds, *Strings, 90'*, World Scientific Publishing, NJ (1991), 401–429.

[9] P. Candelas, X.C. de la Ossa, P.S. Green, and L. Parkes, A pair of Calabi-Yau manifolds as an exactly soluble superconformal field theory, *Nuclear Phys* **B 359** (1991), 21–74.

[10] I. Ciocan-Fontanine, Quantum cohomology of flag manifolds, preprint (1995).

[11] S. Donaldson, The orientation of Yang-Mills moduli spaces and 4-manifold topology, *J. Diff. Geom.* **26** (1987), 397–428.

[12] S. Donaldson, Gluing techniques in the cohomology of moduli spaces, *Topological methods in modern mathematics* (1991), 137–170.

[13] S. Donaldson and P. Kronheimer, *The Geometry of Four-Manifolds*, Oxford University Press, (1990).

[14] S. Dostoglou and D.A. Salamon, Instanton homology and symplectic fixed points, in *Symplectic Geometry*, edited by D. Salamon LMS Lecture Notes Series **192**, Cambridge University Press, (1993), 57–93.

[15] S. Dostoglou and D.A. Salamon, Cauchy-Riemann operators, self-duality, and the spectral flow, in *First European Congress of Mathematics, Volume I, Invited Lectures (Part 1)*, edited by A. Joseph, F. Mignot, F. Murat, B. Prum, R. Rentschler, Birkhäuser Verlag, Progress in Mathematics, **Vol. 119**, 1994, pp. 511–545.

[16] S. Dostoglou and D.A. Salamon, Self-dual instantons and holomorphic curves, *Annals of Math*, **139** (1994), 581–640.

[17] B. Dubrovin, Integrable systems in topological field theory, *Nucl. Phys. B* **379** (1992), 627–689.

[18] A. Floer, Morse theory for Lagrangian intersections, *J. Diff. Geom.* **28** (1988), 513–547.

[19] A. Floer, The unregularized gradient flow of the symplectic action, *Comm. Pure Appl. Math.* **41** (1988), 775–813.

[20] A. Floer, Symplectic fixed points and holomorphic spheres, *Comm. Math. Phys.* **120** (1989), 575–611.

[21] A. Floer and H. Hofer, Coherent orientations for periodic orbit problems in symplectic geometry, *Math. Zeit.* **212** (1993), 13–38.

[22] A. Floer, H. Hofer, and D.A. Salamon, Transversality in elliptic Morse theory for the symplectic action, to appear in *Duke Math. J.* (1995).

[23] D. Gilbarg and N.S. Trudinger, *Elliptic Partial Differential Equations of the Second Order*, Springer-Verlag, (1983).

[24] A. Givental and B. Kim, Quantum cohomology of flag manifolds and Toda lattices, *Commun. Math. Phys.* **168**, (1995), 609–641.

[25] P. Griffiths and J.Harris, *Introduction to Algebraic Geometry*, Wiley (1978).

[26] M. Gromov, Pseudo holomorphic curves in symplectic manifolds, *Invent. Math.* **82** (1985), 307–347.

[27] Grothendieck, Sur la classification des fibrés holomorphes sur la sphère de Riemann, *Amer. J. Math.* **76**, (1957), 121-138.

[28] V. Guillemin and A. Pollack, *Differential Topology*, Prentice Hall, (1974).

[29] Hartman and Wintner, On the local behavior of solutions of non-parabolic partial differential equations, *Amer Journ. Math.* **75** (1953), 449-476.

[30] H. Hofer, Pseudoholomorphic curves in symplectizations with applications to the Weinstein conjecture in dimension three, *Invent. Math.* **114** (1993), 515–563.

[31] H. Hofer, V. Lizan, and J.C. Sikorav, On genericity for holomorphic curves in 4-dimensional almost-complex manifolds, preprint (1994).

[32] H. Hofer and D. Salamon, Floer homology and Novikov rings, Preprint (1993), to appear in *Gauge theory, Symplectic Geometry, and topology, essays in memory of Andreas Floer*, edited by H. Hofer, C. Taubes, A. Weinstein, and E. Zehnder, Birkhäuser (1995).

[33] F. John, *Partial Differential Equations*, Springer-Verlag, (1984).

[34] S. Kobayashi and K. Nomizu, *Foundations of Differential Geometry*, Wiley, (1963).

[35] M. Kontsevich and Yu. Manin, Gromov-Witten classes, quantum cohomology, and enumerative geometry, *Commun. Math. Phys.* **164** (1994), 525–562.

[36] F. Lalonde and D. McDuff, The Geometry of Symplectic Energy, to appear in *Annals of Math.* **141** (1995), 349–371.

[37] B. Lawson, Minimal varieties in real and complex geometry, Sem Math Sup, Vol **57**, Presses Université de Montréal, (1974).

[38] G. Liu, Associativity of quantum multiplication, Preprint (1994).

[39] G. Liu, On the equivalence of Floer and quantum cohomology, Preprint (1995).

[40] W. Lorek, Orientability of the evaluation map for pseudo-holomorphic curves, preprint (1995).

[41] W. Lorek, Regularity of almost-complex structures and generalized Cauchy–Riemann operators, preprint (1995).

[42] D. McDuff, Examples of symplectic structures, *Invent. Math.* **89** (1987), 13–36.

[43] D. McDuff, Elliptic methods in symplectic geometry, *Bull. A.M.S.* **23** (1990), 311–358.

[44] D. McDuff, The local behaviour of holomorphic curves in almost complex 4-manifolds, *Journ. Diff. Geo.* **34** (1991), 143–164.

[45] D. McDuff, The structure of rational and ruled symplectic 4-manifolds, *Journ. Amer. Math. Soc.* **3** (1990), 679–712; Erratum: *Journ. Amer. Math. Soc* **5** (1992), 987–988.

[46] D. McDuff, Symplectic manifolds with contact-type boundaries, *Invent. Math* **103**, (1991), 651–671.

[47] D. McDuff, Immersed spheres in symplectic 4-manifolds, *Annal. de l'Inst Fourier*, **42**, (1991), 369–392.

[48] D. McDuff, Singularities of *J*-holomorphic curves in almost complex 4-manifolds, *Journ. Geom. Anal.* **3**(1992), 249–266.

[49] D. McDuff, Notes on Ruled Symplectic 4-manifolds, *Trans. Amer. Math. Soc.* **345** (1994), 623–639.

[50] D. McDuff, Singularities and Positivity of Intersections of *J*-holomorphic curves, with Appendix by Gang Liu, in [5], 191–216.

[51] D. McDuff, *J*-holomorphic spheres in symplectic 4-manifolds: a survey, to appear in *Proceedings of the Symposium on Symplectic Geometry*, held at Isaac Newton Institute 1994, LMS Lecture Note Series, Cambridge University Press.

[52] D. McDuff and D.A. Salamon, *Introduction to Symplectic Topology*, Oxford University Press, (1995).

[53] M. Micallef and B. White, The structure of branch points in area minimizing surfaces and in pseudo-holomorphic curves, *Annals of Math.* **141** (1995), 35–85.

[54] J.W. Milnor, *Topology from the differential viewpoint*, The University Press of Virginia, (1965).

[55] J.W. Milnor and J.D. Stasheff, *Characteristic Classes*, Princeton University Press, (1974).

[56] J. Moser, Finitely many mass points on the line under the influence of an exponential potential - an integrable system, in *Dynamical Systems, Theory and Applications*, Springer-Verlag, New York, (1975), 467–497.

[57] A. Nijenhuis and W. Woolf, Some integration problems in almost-complex and complex manifolds, *Annals of Math.* **77** (1963), 424–489.

[58] S.P. Novikov, Multivalued functions and functionals - an analogue of the Morse theory, *Soviet Math. Dokl.* **24** (1981), 222–225.

[59] Y.-G. Oh, Removal of Boundary Singularities of Pseudo-holomorphic curves with Lagrangian boundary conditions, *Comm. Pure Appl. Math.* **45** (1992), 121–139.

[60] K. Ono, The Arnold conjecture for weakly monotone symplectic manifolds, *Invent. Math.* **119** (1995), 519–537.

[61] P. Pansu, Pseudo-holomorphic curves in symplectic manifolds, Preprint (1986), in [5], 233–250.

[62] T.H. Parker and J.G. Wolfson, Pseudoholomorphic maps and bubble trees, *Journ. Geom. Anal.* **3** (1993), 63–98.

[63] S. Piunikhin, D. Salamon, M. Schwartz Symplectic Floer-Donaldson theory and quantum cohomology, Preprint, University of Warwick (January 1995), to appear in *Proceedings of the Symposium on Symplectic Geometry*, held at Isaac Newton Institute 1994, ed. C. Thomas, LMS Lecture Note Series, Cambridge University Press.

[64] Y. Ruan, Topological Sigma Model and Donaldson type invariants in Gromov theory, Preprint (1993).

[65] Y. Ruan, Symplectic Topology on Algebraic 3-folds, *Journ. Diff Geom.* **39** (1994), 215–227.

[66] Y. Ruan, Symplectic topology and complex surfaces, in *Geometry and Topology on complex manifolds*, ed. T. Mabuchi, J. Nogichi, T. Ochial, World Scientific Publications (1994).

[67] Y. Ruan and G. Tian, A mathematical theory of quantum cohomology (announcement), *Math. Res. Letters* **1** (1994), 269–278.

[68] Y. Ruan and G. Tian, A mathematical theory of quantum cohomology, Preprint (1994), to appear in *J. Diff. Geom.*.

[69] Y. Ruan and G. Tian, Bott-type symplectic Floer cohomology and its multiplicative structures, Preprint (1994).

[70] Y. Ruan and G. Tian, Higher genus symplectic invariants and sigma model coupled with gravity, Preprint (1995).

[71] J. Sacks and K. Uhlenbeck, The existence of minimal immersions of 2-spheres, *Annals of Math.* **113** (1981), 1–24.

[72] V. Sadov, On equivalence of Floer's and quantum cohomology, preprint, Harvard University HUTP-93/A027 (1993)

[73] D. Salamon, Morse theory, the Conley index and Floer homology, *Bulletin L.M.S.* **22** (1990), 113–140.

[74] D. Salamon, Quantum cohomology and the Atiyah-Floer conjecture, in preparation.

[75] D. Salamon and E. Zehnder, Morse theory for periodic solutions of Hamiltonian systems and the Maslov index, *Comm. Pure Appl. Math.* **45** (1992), 1303–1360.

[76] M. Schwarz, *Morse Homology*, Birkhäuser Verlag, (1993).

[77] M. Schwarz, PhD thesis, ETH-Zürich, (1995).

[78] M. Schwarz, In preparation.

[79] B. Siebert and G. Tian, On quantum cohomology rings of Fano manifolds and a formula of Vafa and Intriligator, Preprint (1994).

[80] S. Smale, An infinite dimensional version of Sard's theorem, *Am. J. Math.* **87** (1973), 213–221.

[81] C. Taubes, Personal communication.

[82] C. Vafa, Topological mirrors and Quantum rings, in *Essays on Mirror Manifolds*, edited by S.-T. Yau, International Press, Hong Kong (1992).

[83] C. Viterbo, The cup-product on the Thom- Smale-Witten complex, and Floer cohomology, preprint (1993)

[84] P.M.H. Wilson, The Kähler cone on Calabi-Yau threefolds, *Invent. Math.* **107** (1992).

[85] E. Witten, Two dimensional gravity and intersection theory on moduli space, *Surveys in Diff. Geom.* **1** (1991), 243–310.

[86] E. Witten, Supersymmetry and Morse theory, *J. Diff. Geom.* **17** (1982), 661–692.

[87] E. Witten, The Verlinde algebra and the cohomology of the Grassmannian, Preprint, Institute of Advanced Study, Princeton, December (1993).

[88] R. Ye, Gromov's compactness theorem for pseudoholomorphic curves, *Trans. Amer. Math. Soc.* **342** (1994), 671–694.

Index

a priori estimate
    for energy density, 44
action
    Morse-Novikov theory for, 154
    of short loop, 45
    on loop space, 153
adjunction formula, 39, 101
almost complex structure
    K-positive, 59
    K-semi-positive, 59
    ω-compatible, 25
    ω-tame, 2, 42
    condition to be regular, 38
    generic, 5
    good, 98
    integrable, 3
    positive, 59
    regular, 24
    semi-positive, 59, 143
    smooth homotopy, 25
    tame versus compatible, 26, 60
Arnold conjecture, 158, 165
Aronszajn, 15, 197
Aronszajn's theorem, 14, 35
Aspinwall, ix, 107, 151, 197
Aspinwall-Morrison formula, 151
Astashkevich, 119, 197
Atiyah, 197
Atiyah-Floer conjecture, 163, 165
Audin, ix, 1, 197

Bertram, 123, 197
bubbling, 41, 46–49, 155

Calabi-Yau manifold, 1, 12, 137, 141, 145, 148–151
    examples, 148, 149
Calderon-Zygmund inequality, 186–189
Callaghan, 163, 197

Candelas, 141, 148, 197
Cauchy-Riemann equation, 2, 5, 13–15, 190
    perturbed, 118, 144, 193
Cauchy-Riemann operator, 24
    fundamental solution, 190
Chern character, 128
Chern class, 2
    quantum, 121
Chern number, 38
    minimal, 10, 61
    negative, 86, 165
    zero, 148
Ciocan-Fontanine, 119, 197
cohomology
    quantum, 108
compactness, 1, 5–6
compactness theorem, 51, 63, 65, 68
    proof, 52, 81–87
composition rule, 137
condition
    finiteness, *see* finiteness condtion
    $(H1)$, 93, 101
    $(H2)$, 93, 96
    $(H3)$, 96
    $(H4)$, 101, 103, 114
    $(H5)$, 101, 103, 114
    (JA_3), 67
    (JA_4), 68, 86, 142
    (JA_p), 67–69, 85–87, 101
    on dimension, *see* dimension condition
conformal map, 13, 41
conformal rescaling, 41
Conley, 155
convergence problem, 137, 139, 141, 149, 151
critical point, 15
    finite number of, 15

stable and unstable manifolds, 161
cross-ratio, 105
cup product
    deformed, 110, 147
        weakly monotone case, 141
    given by triple intersection, 107
    pair-of-pants, 160
curve, 3–4
    approximate J-holomorphic, 30
    as symplectic submanifold, 3
    counting discrete curves, 101, 138,
        151
    critical point, 15–18
    cusp, *see* cusp-curve
    determined by ∞-jet, 14
    energy, 6
    graph of, 142
    implicit function theorem for, 30
    injective point, 4, 18
    intersections of, 17
    isolated, 89
    multiply-covered, 4, 18, 119, 142–
        144, 165
    of negative Chern number, 86, 165
    parametrized, 3
    positivity of intersections, 21, 39,
        100, 101
    reducible, 42
    regular, 24, 38
    simple, 4, 18–21
    singular point of sequence, 53
    somewhere injective, 4, 18
        is simple, 18
    weak convergence, 50
cusp-curve, 42, 50, 62
    framing D, 63, 76
    label T, 65, 81
    moduli space of, 75
    type D, 76
    with two components, 80

D_u
    definition, 24
    ellipticity of, 28
    formula for, 28
    its adjoint D_u^*, 29
    right inverse, 29, 173–176
    surjectivity of, 29, 38

Darboux's theorem, 2
Daskalopoulos, 123, 197
deformed cup product, *see* cup prod-
    uct
determinant bundle, 33
dimension condition
    for Φ, 93
    for Ψ, 102
    for $a * b$, 110
    for $a \star b \star c$, 114
Donaldson, 33, 162, 167, 168, 197
Donaldson's quantum category, 162–
    165
    for Lagrangians, 164
    for mapping tori, 163
Dostoglou, 30, 155, 162, 163, 197
Dubrovin, 107, 113, 131, 139, 198
Dubrovin connection, 113, 132
    and quantum products, 133
    explicit formula, 134
    flatness, 136
    potential function, 135

elliptic bootstrapping, 41, 43, 192–196
elliptic regularity, 25–27, 29, 35, 191,
    192
energy, 6, 42, 44
    bounds derivative, 44
    conformal invariance, 41
    identity, 41, 43
energy level, 145
evaluation map, 6–8, 49
    and orientations, 98
    as pseudo-cycle, 94–98
    domain of, 64
    for cusp-curves, 79–81
    for marked curves, 66–69
    image of, 62–65
    is submersion, 71–74
    p-fold, 8, 65, 66
        compactification, 8, 65

finiteness condition, 144, 147, 156
finiteness result, 155
first Chern class, 2
flag manifold, *see* quantum cohomol-
    ogy

Floer, ix, 12, 15, 41, 153, 155–158, 162, 164, 198
Floer (co)homology, 1, 11, 12, 153–159
  and quantum cohomology, 160
  cochain complex, 155
  Euler characteristic, 164
  of (H, J), 156
  of symplectomorphism, 162
  pair-of-pants product, 160
  ring structure, 159–160
Floer-Donaldson theory, 163
framing
  D for cusp-curve, 63, 76
Fredholm operator, 5
  determinant bundle, 33
  index, 5, 24
  regular value, 5, 33
Fredholm theory, 1, 4, 5
Frobenius algebra, 113, 131
Frobenius condition, 133

Gilbarg, 186, 188, 198
Givental, ix, 11, 107, 119, 121, 198
Green, 141, 148, 197
Griffiths, 33, 38, 100, 198
Gromov, ix, 7, 41, 50, 89, 107, 143, 198
Gromov invariant Φ, 66, 89, 93–94
  dimension condition, 93
  for 6-manifolds, 100
  for conics in $\mathbb{C}P^2$, 99
  for discrete curves, 101
  for lines in $\mathbb{C}P^n$, 98
  on blow up, 100
Gromov's compactness theorem, 50–58
Gromov-Witten invariant Ψ, 66, 69, 101–105
  dimension condition, 102
Gromov-Witten invariants, 1, 8–9
  compared on $\mathbb{C}P^2$, 104
  composition rule, 117
    on $\mathbb{C}P^2$, 119
  deformation invariance, 96
  mixed, 90, 138
  signs, 98
  weakly monotone case, 141

  well defined for Kähler manifolds, 96
Gromov-Witten potential, 131–139
  for $\mathbb{C}P^2$, 138
  for $\mathbb{C}P^n$, 138
  non-monotone case, 139
Grothendieck, 38, 198
Guillemin, 92, 198

Hölder inequality, 169
Hölder norm, 183
Hamiltonian differential equation, 153
harmonic function, 185
  mean value property, 185
Harris, 33, 38, 100, 198
Hartman, 15, 198
Heegard splitting, 165
Hofer, 12, 15, 40, 41, 51, 54, 146, 153, 155, 157, 158, 162, 198
homology class
  J-effective, 67
  framed, 63
  indecomposable, 49
  of pure degree, 108, 131
  spherical, 6, 47
Hurewicz homomorphism, 6

implicit function theorem, 27–33
integrable systems, 113, 121
isoperimetric inequality, 41, 45

(J, J')-holomorphic, 2
J-holomorphic curve, *see* curve
J-holomorphic map, 13
John, 185, 186, 199

Kähler manifold, 61
  and symplectic deformations, 100
  Fano variety, 60
  Gromov-Witten invariants, 96
Kim, 11, 107, 119, 121, 198
Kobayashi, 199
Kobayaski, 28
Kodaira vanishing theorem, 38
Kontsevich, 90, 113, 131, 137–139, 199
Kronheimer, 167, 168, 197

labelling
  T for cusp-curve, 65, 81

Lafontaine, ix, 1, 197
Lalonde, ix, 7, 199
Landau-Ginzburg potential, 126, 127
Laplace's equation
    fundamental solution, 185
Laurent polynomial ring, 10
Lawson, 44, 199
Liu, 118, 199
Lizan, 40
loop
    action of, 45
loop space
    of manifold, 146, 154
    universal cover, 154
Lorek, 40, 98, 199

Manin, 90, 113, 131, 137–139, 199
mapping tori
    and Atiyah-Floer conjecture, 163
marked sphere, 105
Maslov index, 155
mass
    of singular point, 53
McDuff, ix, 1, 2, 7, 17, 21, 28, 33, 39,
       40, 59, 62, 74, 77, 89, 98–
       101, 156, 199, 200
Micallef, 21, 200
Milnor, 92, 129, 200
minimal Chern number, 10, 61
mirror symmetry conjecture, 122, 148,
       149
moduli space, 4–5
    cobordism of, 25
    compactifications, 59–69
    complex structure on, 33
    integration over, 97, 103
    main theorems, 23–25
    of N-tuples of curves, 74–75
    of cusp-curves, 75–78
    of flat connections, 163
    of two-component cusp-curves, 80
    of unparametrized curves, 52
    orientation, 32, 33, 98
    regular point of, 24
    top strata in the boundary, 82
    universal, 33, 77
    Yang-Mills, 33
monotone, *see* symplectic manifold

Morrison, ix, 107, 151, 197
Morse-Novikov theory, 146, 154
Morse-Witten coboundary, 158
Morse-Witten complex , 161
Moser, 121, 122, 200

Nijenhuis, 74, 200
Nijenhuis tensor, 28
    anti-commutes with J, 32
Nomizu, 28, 199
non-squeezing theorem, 7
Novikov, 146, 154, 200
Novikov ring, 11, 12, 141, 144–148
    as Laurent series ring, 146
number of curves
    in $\mathbb{C}P^2$, 105, 138
    in Calabi-Yau manifold, 101, 151

Oh, 200
Ono, 153, 158, 200
Ossa, 141, 148, 197

Pansu, 41, 44, 51, 200
Parker, 41, 51, 200
Parkes, 141, 148, 197
Piunikhin, 157, 161, 200
Poisson's identity, 185
Pollack, 92, 198
potential function, 135
product, *see* cup product
pseudo-cycle, 8, 63, 66, 68, 90–93
    bordant, 90
    of dimension k, 90
    transverse, 91
    weak representative, 93

quantum
    category, 162, 164
    Chern classes, 121
quantum cohomology, 1, 9–11, 108–
       110
    and Floer cohomology, 157, 160
    as Lagrangian variety, 122
    cup product, 110–114, 147
        associativity, 114–119, 161
        dimension condition, 110
    Frobenius structure of, 113, 131
    of $\mathbb{C}P^n$, 11, 113

of flag manifold, 11, 110, 119–122, 146
of Grassmannian, 123–130
periodic form, 12, 147
weakly monotone case, 148
with Novikov rings, 146

rational maps Rat_m of $\mathbb{C}P^1$, 83, 150
regular
  J-holomorphic curve, 24, 39
  value of Fredholm operator, 33
Rellich's theorem, 27, 184
removal of singularities, 41, 43–46
reparametrization group G, 42, 62
Riemann-Roch theorem, 24, 33
Ruan, ix, 9, 33, 50, 59, 62, 64, 66, 76, 89, 90, 100, 101, 104, 105, 107, 117, 118, 131, 137, 138, 144, 200, 201

Sacks, 41, 201
Sadov, 119, 161, 197, 201
Salamon, ix, 1, 2, 7, 12, 15, 30, 41, 44, 51, 100, 146, 153, 155–158, 161–163, 197–201
Sard's theorem, 5, 24
Sard-Smale theorem, 33, 36, 78
Schwarz, 157, 158, 161, 201
semi-positivity, 59
Siebert, 121, 123, 124, 201
sigma model, 89
sign
  convention for Lie bracket, 132
  of intersection, 98
Sikorav, 40
singular point of curve, 53
Smale, 33, 201
smoothness
  of class $W^{k,p}$, 181, 184
smoothness convention, 34
Sobolev embedding theorem, 27, 184
Sobolev estimate
  borderline case, 6, 41, 184
Sobolev space, 181–184
spherical, 47
Stasheff, 129, 200
supermanifold, 131, 139
symplectic action, *see* action

symplectic form
  deformation, 96
symplectic manifold, 1–3
  monotone, 1, 8, 60
  non-equivalent, 100
  weakly monotone, 59–61, 141, 153
  Arnold conjecture for, 158

Taubes, ix, 33, 36, 201
Tian, ix, 90, 104, 107, 117, 118, 121, 123, 124, 131, 137, 138, 144, 201
Toda lattice, 119, 121
triple intersection product, 107
Trudinger, 186, 188, 198

Uhlenbeck, 41, 201
unique continuation, 15
universal moduli space, 33, 77
  is a manifold, 34, 77

Vafa, ix, 107, 201
Verlinde algebra, 123, 128–130
  gluing rules, 130
Viterbo, 160, 201

WDVV-equation, 11, 135, 137
weak convergence
  definition, 50
weak derivative, 181
weak solution, 185, 191
weakly monotone manifold, *see* symplectic manifold
Wentworth, 123, 197
Weyl's lemma, 186
White, 21, 200
Wintner, 15, 198
Witten, 89, 107, 123, 124, 130, 158, 202
Wolfson, 41, 51, 200
Woolf, 74, 200

Yau, 148
Ye, 41, 51, 202

Zehnder, 155–157, 201

Index of Notations

$\mathcal{J}_\tau(M,\omega)$ 2

$c_1(A)$ 2

$u : (\Sigma, j) \to (M, J)$ 3

$C = \operatorname{Im} u$ 3

$\mathcal{M}(A, J)$ 4, 23

$\mathcal{J}_{\mathrm{reg}}(A)$ 4, 24

$\langle v, w \rangle$ 5

$G = \operatorname{PSL}(2, \mathbb{C})$ 6

$\mathcal{W}(A, J) = \mathcal{M}(A, J) \times_G \mathbb{C}P^1$ 7, 62

$e_J : \mathcal{W}(A, J) \to M$ 7

$\Phi_A(\alpha_1, \ldots, \alpha_p)$ 9, 94

$\Psi_A(\alpha_1, \ldots, \alpha_p)$ 9, 99

$\alpha = \operatorname{PD}(a)$ 10, 106

$a * b$ 10, 108

$QH^*(M)$ 10, 106

$\widetilde{QH}^*(M) = H^*(M) \otimes \mathbb{Z}[q]$ 10, 107

$\Gamma = \operatorname{Im} \pi_2(M) \to H_2(M)$ 11, 142

Λ_ω 11, 142

$\bar{\partial}_J$ 13

$\mathcal{X} = \operatorname{Map}(\Sigma, M; A)$ 23

D_u 24, 28

$\mathcal{J}_{\mathrm{reg}}(A)$ 24

$\mathcal{J}(M, \omega)$ 24

$(u, J) \in \mathcal{M}^\ell(A, \mathcal{J}) \subset \mathcal{X}^{k,p} \times \mathcal{J}^\ell$ 34

$\operatorname{End}(TM, J, \omega)$ 34

$\pi : \mathcal{M}^\ell(A, \mathcal{J}) \to \mathcal{J}^\ell$ 36

$g_J(v, w) = \langle v, w \rangle_J$ 42

$E(u; B_r)$ 44

$C = C^1 \cup C^2 \cup \ldots \cup C^N$ 50

$u = (u^1, \ldots, u^N)$ 50

$A(r, R)$ 52

$\mathcal{J}_+(M, \omega), \mathcal{J}_+(M, \omega, K)$ 59

$D = \{A^1, \ldots, A^N, j_2, \ldots, j_N\}$ 63

$\mathcal{W}(D, J)$ 64, 79

$e_p : \mathcal{W}(A, J, p) \to M^p$ 65

$\mathbf{z} = (z_1, \ldots, z_p)$ 66

$e_{\mathbf{z}} = e_{A, J, \mathbf{z}} : \mathcal{M}(A, J) \to M^p$ 66

$e_{D, T, \mathbf{z}} : \mathcal{V}(D, T, J, \mathbf{z}) \to M^p$ 68

$\mathcal{M}(A^1, \ldots, A^N, \mathcal{J})$ 74

Δ_N 76

$\pi_D : \mathcal{M}(D, \mathcal{J}) \to \mathcal{J}$ 78

$\Phi_{A,p} : H_d(M^p, \mathbb{Z}) \to \mathbb{Z}$ 93

$\Psi_{A,p} : H_d(M^p, \mathbb{Z}) \to \mathbb{Z}$ 102

$\langle a, b \rangle$ 108, 131

$(a * b)_A$ 111

$\Psi_{A,B}(\alpha, \beta; \gamma, \delta)$ 116

$\widehat{J}, \widehat{A}$ 143

$W^{k,p}(\Omega), W_0^{k,p}(\Omega)$ 181